THE CITY OF TOMORROW

SENSORS,

NETWORKS,

HACKERS,

THE CITY OF TOMORROW

AND THE

FUTURE OF

URBAN LIFE

CARLO RATTI AND

MATTHEW CLAUDEL

Yale UNIVERSITY PRESS   NEW HAVEN AND LONDON

Yale University Press books may be purchased in quantity for educational, business, or promotional use. For information, please e-mail sales.press@yale.edu (U.S. office) or sales@yaleup.co.uk (U.K. office).

Designed by Nancy Ovedovitz and set in Scala and The Sans types by Integrated Publishing Solutions. Printed in the United States of America.

Library of Congress Control Number: 2015955547
ISBN 978-0-300-20480-3 (hardcover : alk. paper)

A catalogue record for this book is available from the British Library.

This paper meets the requirements of ANSI/NISO Z39.48–1992 (Permanence of Paper).

10 9 8 7 6 5 4 3 2 1

Frontispiece: *New York Talk Exchange* (detail; see chapter 2)

CONTENTS

PART I THE CITY OF TOMORROW (AND TODAY)

We are called to be architects of
the future, not its victims.
R. Buckminster Fuller, 1969

ONE
FUTURECRAFT

On December 24, 1900, the *Boston Globe* ran a piece imagining what Boston would look like at the turn of the millennium. The lavishly illustrated article by Thomas F. Anderson painted an elaborate vision of a city with moving sidewalks, airships soaring high above the streets, and pneumatic tube delivery of everything from newspapers to food. The author's predictions were sweeping and optimistic: Boston would be so beautiful that the word "slum" would be eliminated from the city's vernacular.[1]

Such descriptions, in retrospect, are almost comical—yet a hope of glimpsing the future continues to enchant us. A vibrant thread in fiction and film, speculations have become a genre in their own right, encompassing, as a common trope, the future of the city. Notions vary widely, from H. G. Wells' grim dystopias to Fritz Lang's *Metropolis* or the pseudo-police state of *Minority Report*. Regardless of the time and medium, however, "nothing ever looks as dated as old science fiction," as the saying goes. Futures quickly become *paleofutures*—early speculations about the futures that would never come to pass.

In the midst of a sprawling graveyard of ideas, an exercise like this book—an exploration of cities as part of Yale University Press's series on the future—is vexed with a crucial question: Is it possible for our predictions to escape the fate of Anderson's? How can we avoid the scrap heap of urban visions? And, more specifically, Does the act of considering the future—in this case, the future of the city—have inherent and productive value?

Traditionally, most future visions have been attempts to accurately depict the world of tomorrow—and that may be their undoing. Prediction often involves assaying weak signals at the cutting edge of the contemporary world and flinging them far forward, for decades or centuries, to arrive at a portrait of the future city. To Anderson, writing in the year 1900, soon after the dazzling introduction of zeppelin travel and pneumatic technology, it seemed all but assured that these advances would define urban development over the course of the next hundred years. The state of the art stirred his imagination and defined his portrait of millennial Boston.

We propose something quite different: to employ design in a systematic exploration and germination of possible futures. Our aim is not to portray what is to come. Rather, we apply a method that we call *futurecraft:* we posit future scenarios (typically phrased as What if? questions), entertain their consequences and exigencies, and share the resulting ideas widely, to enable public conversation and debate. In other terms, we propose to extrapolate from the present condition and to place ourselves, as designers, in a fictive but possible future context with the intent of realizing or precluding that future through public discourse.

This concept, primarily developed through our research at the Massachusetts Institute of Technology's Senseable City Lab, has antecedents. Recently, Anthony Dunne and Fiona Raby at the Royal College of Art in London proposed "speculative design"—a process that acts as a "catalyst for collectively redefining our relationship to reality" and considering how things could be. An earlier framework, Comprehensive Anticipatory Design Science (CADS) proposed by the iconic inventor Buckminster Fuller, uses a systematic approach to design that he developed through a class at MIT in 1956. Motivating Fuller in his work was a general belief that design, speculation, and science go hand in hand. "The function of what I call *design science* is to solve problems by introducing into the environment new artifacts, the availability of which will induce their spontaneous employment by humans and thus, coincidentally, cause humans to abandon their previous problem-producing behaviors and devices."[2]

Buckminster Fuller's statement suggests a latent evolutionary concept. As technical culture progresses, objects are produced and iteratively refined through design—an act that introduces mutations to improve a function or enable a new capability. On a broad scale, these mutations collectively promote change and development. In an 1863 text, "Darwin among the Machines," the writer Samuel Butler proposed an evolutionary analogy for technology: replacing organisms with artifacts and classifying the synthetic kingdom into genera and species.[3]

Variations of this concept have recurred both in theory and in practice.[4] If we accept this evolutionary framework, a central question emerges: How can the designer accelerate positive technological change? To continue the biological analogy, could the crucial role of the designer be to produce anomalies (as new ideas)? The designer could become what, in biology, is referred to as a mutagen—an agent of mutation. Although mutations in the natural world are random, our concept of design is directed by futurecraft.

Most importantly, futurecraft is not about fixing the present (an overwhelming task) or predicting the future (a disappointingly futile activity) but influencing it positively. Designers should not force their ideas into the world—in fact, whether or not an idea is realized is largely irrelevant. By virtue of being stated,

6

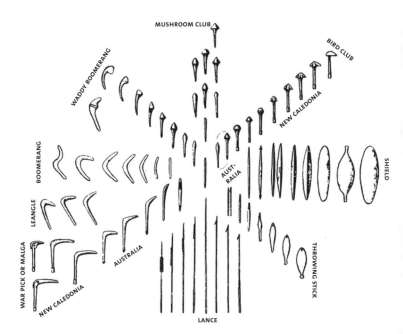

MUSHROOM CLUB

BIRD CLUB

WADDY BOOMERANG

NEW CALEDONIA

BOOMERANG

SHIELD

AUST-
RALIA

LEANGLE

WAR PICK OR MALGA

AUSTRALIA

THROWING STICK

NEW CALEDONIA

LANCE

The Evolution of Culture by Augustus Henry Lane-Fox Pitt-Rivers

According to a long-standing hypothesis, the synthetic world—the kingdom of human-made objects—evolves in a way that is analogous to biological evolution: by iterative generations, small mutations, and natural selection. In 1852, Augustus Henry Lane-Fox Pitt-Rivers, a British imperial officer and avid collector, was commissioned by the Crown to prepare a manual on firearms. In studying the history of weaponry, he became convinced of the gradual technical development of human artifacts over time. This image shows an arrangement of primitive weapons in his collection, which points to what he called "the evolution of culture." He devoted his later life to "evolutionary anthropology," testing the concept of synthetic evolution through his vast collection of artifacts. Although his assumptions are now regarded as contrived, the application of an evolutionary analogy to the synthetic world remains a useful interpretive tool.

7

explored, and debated, a concept will necessarily make an impact. Provocation is a better metric than certainty, for ideas both positive and negative. Effacing a dystopian vision for the sake of decency is a disservice, precluding the possibility of avoiding that future.

Methodologically, futurecraft dissolves prediction anxiety and opens up new avenues of research rather than delivering products and systems. Designers must not, however, peddle only abstract ideas—tangible demonstrations are crucial to promoting general discussion. At the urban scale, these enable interaction with people—future users—and demonstrate ideas that spark development. Specific mutations are tested in urban space and subjected to public debate, a process that functions like natural selection in biology. The public will eventually steer broader technological development toward the most desirable future.

One should note that this process is not limited to areas traditionally identified as leaders of technological progress. In this book, we deliberately focus on new ideas at the cutting edge, where—almost by definition—every original concept begins before spreading to different contexts. This dissemination, particularly in the developing world, or contexts that do not have similar preexisting technologies, can cause a "leapfrogging" effect. Cell phones, for example, have become tremendously widespread across the African continent in only a few years—while

Western countries went through a long, protracted development from analogue landlines. Countries without an existing telecommunications infrastructure could leap directly to the latest technology, bypassing interim stages. While we will not focus here on such developing contexts, we recognize that they could be one of the most fertile grounds for the development of futurecraft.

Cities are, by definition, plural, public, and productive. They are created by society itself (barring exceptional cases like master-planned Brasília or Chandigarh), and they function as culture's petri dish for progress. *Living in* space and *creating* space can go hand in hand. "To achieve change," in the words of Dunne and Raby, "it is necessary to unlock people's imaginations and apply it to all areas of life at a microscale. Critical design, by generating alternatives, can help people construct compasses rather than maps for navigating new sets of values."[5]

Our work is meaningless unless it ignites imaginations and provokes debate: design by mutation is intrinsically *collective*. Designers produce mutations, some of which will grow, evolve, and develop into tangible artifacts that cause global change— driven to realization by the energy of the crowd. Crucially, this process depends on channels for disseminating information from designers to citizens—media, museums, exhibitions, publications. This book itself is a vector for transmission, part of the idea propagation that is integral to futurecraft.

The method and the function of futurecraft are best shown in specific examples. Trash Track, a 2009 project by the Sense-able City Lab, imagined a future scenario in which geolocating devices become so small and inexpensive that almost everything could be tagged. Researchers proposed a design into that scenario—trash that wirelessly reports its GPS location—and created a full-scale urban demonstration to test it. With the help of hundreds of citizen volunteers, the team deployed thousands of sensors into Seattle's waste management system and watched as the tags traced waste movement across the United States. A set of visualizations and videos revealed the inefficiencies of the disposal chain and were communicated widely through exhibitions, news, and other media. Subsequent discussion and debate has led to systemic improvements by waste management companies, inspired startups that produce trash trackers, and, most importantly, sparked behavioral change in citizens who are motivated to reduce their waste and to recycle. Trash Track exemplified a new relationship between designers and the public, demonstrating the power of futurecraft to shape urban development.

It is a fundamental responsibility of design to challenge the status quo, to introduce new possibilities, to materialize aberrations, and ultimately to pave the way for the public to realize a desirable future. Herbert Simon, echoing Albert Einstein, wrote that "sciences are concerned with how things are . . . design on the other hand is concerned with how things ought to be."[6]

A concern with how things ought to be encompasses a variety of designerly endeavors, everything from aesthetic gloss to problem solving. Many roles within this spectrum can serve a valuable purpose. Aesthetics are crucial to marketability, and a problem-solving mentality can identify areas that are lacking and can generate improvements—yet futurecraft is far removed from these approaches. It is situated one step into the future, focused more on what could be than what is. Echoing Cedric Price in his provocation to the architecture profession, we believe design needs a shift in methods and goals: "Like medicine, [design] must move from the curative to the preventive."[7] Our playing field is tomorrow.

The designer is inherently optimistic in that ideas can be a catalyst for positive change. However, the framework of willful synthetic evolution hinges on an explicit and defined relationship to the future, structured by four core ideas: that the articulation of future conditions is a hypothetical tool; that futurecasting is only part of the enterprise meant to enable and provoke design; that possible futures are rooted in the present and not in distant, idealized, extraordinary, or digressive visions, which means balancing provocation with strong ties to the world-as-it-is; and finally, that whether or not the imagined scenarios come to pass is irrelevant. We are well aware that, in all likelihood, the future will look different from our "what-if" snapshots, but designing into a projected situation can nonetheless guide us toward a possible and desirable future.

The downfall of Anderson's vision for Boston was a disjunction between time frame and reality. In his time, moving sidewalks seemed quite probable one hundred years in the future, but over the course of the twentieth century technological evolution branched in wildly different directions. Anderson imagined moving sidewalks; he could not have imagined Uber. The aim of futurecraft is to maximize impact by aligning its scope with its reach.

Two variables—time and topic—govern this book and our work as designers. The chronology of each chapter encompasses a loosely defined "near future." As a logical extension of the present, design in this arena is immediate and relevant, with potential to reflexively influence today's urban evolution. Each chapter will delve into a specific topic, testing and extrapolating trends within the time frame to understand their promise and their consequence.

In this book we address ideas that will shape the form and function of cities in today's world of bits and atoms. Taking a human-centric approach and acknowledging citizens as the crucial actuators of urban development, we examine urban information flows from the macro to the micro scale. We apply data-driven models to a spectrum of urban systems, from transportation to energy to fabrication and learning. Finally, we circle back to citizens themselves: you, all of us, who together constitute the active urban network. *Hack the city!*

We believe that in each of these domains, the future city will grow from a symbiosis between design and the public. Where these worlds intersect, we can collectively imagine, examine, choose, and create the most desirable future. We ask you, the interested reader, to take this book as a collection of mutations that can spark debate and open new lines of research. Even if they do not materialize, they will have tested the future and steered technological development. "The universe of possible worlds is constantly expanding and diversifying," wrote Lubomír Doležel in *Heterocosmica*, "thanks to the incessant world-constructing activity of human minds and hands . . . the most active experimental laboratory of the world-constructing enterprise."[8] Using futurecraft, we seek to outline possible scenarios, test them in urban space, and let them propagate, ultimately accelerating urban evolution. Design can become an operative mechanism for crowdsourcing the future based on mutation and selection. By soliciting ideas, response, and action from citizens we hope that design can move society toward the most desirable outcome, a *futur souhaité*. The iconic adage from the computer scientist Alan Kay rings true: "The best way to predict the future is to invent it."[9]

*Ubiquitous computing names
the third wave in computing,
just now beginning. First were
mainframes, each shared by lots
of people. Now we are in the personal computing era, person and
machine staring uneasily at each
other across the desktop. Next
comes ubiquitous computing,
or the age of calm technology,
when technology recedes into the
background of our lives.*
Mark Weiser, 1996

TWO
BITS AND
ATOMS

A new form of communication exploded into the early twentieth century, wildly skewing the nature of human connectivity with a sudden force: mass media. The way humans have always related—face-to-face dialogue between neighbors and friends—was expanded by orders of magnitude. With this amplification, elements of the village, whether social or functional, took on new reactive properties, and the world shrank dramatically. Marshall McLuhan, one of the fathers of social media theory, described the universal connective paradigm as a global village: an entire planet of people living as neighbors, suddenly given the tools to speak, or shout, around the world. Humanity was connected from any and every location.

Yet in McLuhan's time the idea of the global village accounted only for unidirectional mass media like radio and television. Information streamed outward, from privileged content-creators to distributors to passive consumers. Universal communication functioned more as a megaphone than as a telephone, amplifying inherent tensions in society rather than promoting cohesion. McLuhan readily acknowledged that "the more you create village conditions," the more you generate "discontinuity and division and diversity. The Global Village absolutely insures maximal disagreement on all points. It never occurred to me that uniformity and tranquility were properties of the Global Village. It has more spite and envy. The spaces and times are pulled out from between people. A world in which people encounter each other in depth all the time. The tribal-global village is far more divisive—full of fighting—than any nationalism ever was. Village is fission, not fusion, in depth all the time."[1] Unidirectional mass media brought a clash of polemics on the global scale.

In the 1980s, soon after McLuhan died, a new connective infrastructure arose that would cause even more sweeping and dramatic changes. The bidirectional connective interface of the Internet became a jumble of top-down and bottom-up energy. More than could ever have been possible through television or radio, people began to share ideas, thoughts, work, obsessions, and intimacies to the widest extent of the network. The choke points of media providers were opened (though not obliterated), and content was democratized to a certain extent. Media became dialogue rather than monologue, and it was at this moment that humanity began coming together as a real village, with shared culture, ideas, and discussion.

People were unified by a pervasive "space of flows." "There is a new spatial form characteristic of social practices that dominate and shape the network society: the space of flows," wrote Manuel Castells, the sociologist who coined the term. "The space of flows is the material organization of time-sharing social practices that work through flows. By flows I understand purposeful, repetitive, programmable sequences of exchange and interaction between physically disjointed positions held by social actors."[2] That is, physical space can no longer be considered absolute. It cannot be divorced from its digital dimension.

Neither could this new system be neutral. The space of flows refers to a merger of virtual networks and material space—one in which digital and physical configurations actively influence one another. But how? What effect would the space of flows have on the physical city? In the looming shadow of the ubiquitous Internet, would the specificity of place have any significance?

A prevailing opinion at this crucial moment in human's cul-

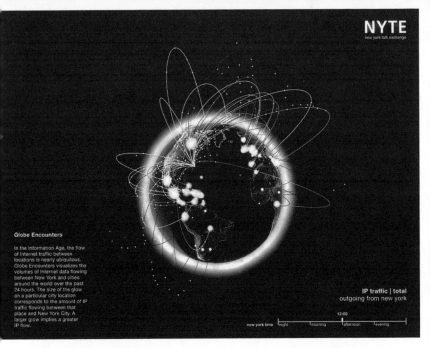

Globe Encounters

In the Information Age, the flow of Internet traffic between locations is nearly ubiquitous. Globe Encounters visualizes the volumes of Internet data flowing between New York and cities around the world over the past 24 hours. The size of the glow on a particular city location corresponds to the amount of IP traffic flowing between that place and New York City. A larger glow implies a greater IP flow.

IP traffic | total
outgoing from new york

new york time | night | morning | 12:00 afternoon | evening

New York Talk Exchange by the MIT Senseable City Laboratory

According to Manuel Castells, the space of flows—transfers of digital information across and throughout space—cannot be separated from our physical world. The contemporary condition is animated by "sequences of exchange and interaction between physically disjointed positions held by social actors." This image—from a project called New York Talk Exchange by the Senseable City Lab—reveals the relationships that New Yorkers share with the rest of the world through a visualization of long-distance telephone calls and Internet data flowing into and out of the city. How does New York connect to other urban spaces? With which cities does New York have the strongest ties, and how do these relationships shift with time? How does the rest of the world reach into the neighborhoods of New York? First exhibited at the Museum of Modern Art in 2008, *New York Talk Exchange* shows the human story hiding within the global space of flows.

tural history was that distance would die. Physicality, it seemed, would lose all relevance as it was subsumed by the connective fabric of the Internet.

The argument held that if information can be instantaneously transferred anywhere, to anyone, then all places are equivalent. If I am connected, why does it matter where I am? "The post-information age will remove the limitations of geography. Digital living will include less and less dependence upon being in a specific place at a specific time, and the transmission of place itself will start to become possible," wrote the MIT Media Lab founder Nicholas Negroponte.[3] Work is a simple example: why commute to the office when the office will come right to your home?

The Internet was expected to neuter place in every dimension of human habitation, from entertainment to employment. Many of the tools for interaction, commerce, and information management were digitized and dematerialized. They became efficient, accessible, and—most significantly—aspatial. The economist Frances Cairncross followed this trend to its logical conclusion with an overt hypothesis that she called the "death of distance." The Internet would usher in a "communications future . . . in which distance is irrelevant."[4]

These are resounding predictions, but history (so far) has proven them wrong. Over the past two decades, cities have grown as never before. Urban space has flourished across the globe as humanity rushes headlong into an urban era. Some calculations suggest that the urban population is increasing by a quarter million *per day,* amounting to a new London every month.[5] The year 2008 was a decisive turning point—when more than half of humanity lived in cities—and growth has only accelerated since. Statistics from the World Health Organization suggest that 75 percent of humans might be city dwellers by 2050, and in China alone, the urban population has risen by more than 500 million during the thirty years since economic liberalization—the equivalent of the populations of the United States plus three Britains. Even by conservative estimates, this constitutes the biggest and fastest shift of humanity that the planet has ever seen.[6] More than ever, cities are human magnets.

Why? It seems that in the collective frenzy of the network, the death-of-distance theorists forgot something crucial to human experience: the importance of physical interaction between people and with the environment. *E-topia,* written in 1999 by the architect and academic William Mitchell, was somewhat of a repent. Mitchell, head of the MIT Media Lab's Smart Cities group, illustrated his point with a humorous vignette about a man living at and running a business from the top of a mountain. The man was no less efficient for working at one of the most remote places on earth, but Mitchell concluded, Who could bear to work in that way? This insight is intuitively clear, but it can also be corroborated empirically. Researchers at the Senseable City Lab

analyzed telecommunication data and meetings and found that
people who communicate digitally also tend to meet in person.[7]
People fundamentally want to be with other people, they want
to be in a beautiful place, they want to be at the center of it all:
people want to live in cities.

"Traditional urban patterns cannot coexist with cyberspace.
But long live the new, network-mediated metropolis of the
digital era." Today's reality is a powerful collision of physical
and digital that augments both—a triumph of atoms and bits.
"To pursue this agenda effectively, we must extend the defini-
tions of architecture and urban design to encompass virtual
places as well as physical ones, software as well as hardware."[8]
Rather than the network subsuming and replacing space, the
two are becoming increasingly enmeshed.

In short, the digital revolution did not kill urban spaces—far
from it—but neither did it leave them unaffected. The introduction of the Internet, the space of flows, the connective tissue
that theorists from Cairncross to Negroponte expected to kill
physical proximity, has indeed had a profound impact on cities.
Instead of flows replacing spaces and bits replacing atoms, cities
are now a hybrid space at the intersection of the two. Physical
and virtual are fused through a productive collision, where both
propinquity and connectivity play an important role.

The new domain of digitally integrated urban space has
come to be known as the smart city. Ubiquitous technology is

suffusing every dimension of urban space, transforming it into a computer for living in (paraphrasing Le Corbusier, the early twentieth-century Swiss architect who crystallized the spirit of his time with his iconic concept of machines for living in).[9] The new city is a fundamentally different space—one where digital systems have a very real impact on how we experience, navigate and socialize.

What is happening at an urban scale today is similar to what happened two decades ago in Formula One auto racing.[10] Up to that point, success on the circuit was primarily credited to a car's mechanics and the driver's capabilities. But telemetry technology completely changed the competition. Digital systems allowed a vehicle to communicate instantaneously with its ground crew. The car became a "computer on wheels," monitored by thousands of sensors and optimized for performance in real time. Formula One winners now work together with "intelligent" vehicles that sense and respond with lightning precision to the conditions of the race. Today, winning is as much about the team behind the computers as it is about the driver behind the wheel.

The Formula One car effectively became a real-time control system—a loop with both sensing and actuating components. The sensors give constant information about conditions and performance, and the actuators can, in turn, have an effect on performance. As sensors and actuators inform each other, they

Formula One Telemetry of Adrian Sutil

Telemetry technology—wireless, nearly instant data transmission—has completely transformed Formula One automobile racing. Today, the frame, the engine, and the mechanisms of the car—even the driver behind the wheel—are only part of the picture. An array of sensors embedded in the vehicle constantly monitors everything from tire pressure to engine temperature. That information is transferred back to a ground crew, who use real-time analytics to optimize performance and immediately respond to race conditions.

Shown here is Adrian Sutil's pit crew during the qualifying session of the Gran Premio de España, May 9, 2009. Just as digital technologies have changed Formula One racing by integrating sensors and actuators in cars, they are quickly transforming our cities. Real-time control systems may monitor and respond at the urban scale, working with citizens and dynamically reacting to the environment or the population's behavior.

work together toward optimizing the system. In the case of the Formula One race car, they might deal with weather conditions and acceleration in a curve—but something similar can now happen in cities.

Already in 2001, a report from the National Academy of Sciences noted that "networks comprising thousands or millions of sensors could monitor the environment, the battle-field, or the factory floor; smart spaces containing hundreds of smart surfaces and intelligent appliances could provide access to computational resources."[11] In smart cities, an ecosystem of sensors collects information from urban space, and an array of network-enabled actuators can subsequently transform that space. Data-driven feedback loops turn the city into a reflexive test-bed and workshop for connected habitation in enmeshed digital and physical space, with a common platform of ubiquitous computing. Within the field of smart cities, a plethora of approaches to theory and practice have emerged, addressing crucial topics from civic hacking to data management to programmable architecture and even autonomous, sentient space.

All of this has repercussions in digital space: almost every contemporary action and interaction creates data. Broadband fiber-optic and wireless telecommunications grids are supporting mobile phones, smartphones, and tablets that are increasingly affordable. At the same time, open databases—informal collaborations between citizens and governments—are aggregating and revealing all kinds of information. The resulting profusion of urban big data opens a fertile ground for research, theory, and practice. What could only be inferred from basic surveys or expensive observational studies during the analogue era can now be immediately "sensed" on a tremendous scale. From social science to mathematics to economics, we can now use these data to address deep questions about how humanity lives. Citizens are empowered to think, act, and transform their public space; they are creating a groundswell of urban innovation that is only just rising today. We are witnessing a "reorientation of knowledge and power" in the city as profound as the transformations that the anthropologist Christopher Kelty has described in the virtual world.[12] This is a new era for the global village: an Internet-mediated space of communication and habitation.

The city is humanity's laboratory,
where people flock to dream,
create, build, and rebuild.
Edward Glaeser, 2012

THREE
WIKI CITY

Visions of the future city, beautiful and ticking like clockwork, have fascinated planners for centuries. In 1922, Le Corbusier aimed to achieve a sophisticated feat of urban engineering, a city of optimal efficiency, which he presented with lyrical prose. "I should like to draw a picture of 'the street' as it would appear in a truly up-to-date city . . . You are under the shade of trees, vast lawns spread all round you . . . Look through the charmingly dispersed arabesques of branches out into the sky towards those widely-spaced crystal towers which soar higher than any pinnacle on earth."[1]

Today, almost a century later, we "imagine a late summer afternoon in Songdo a few years from now" with the

urban researcher and theorist Anthony Townsend. "Smart buildings might order millions of remotely controlled motors to open
windows and blinds to catch the evening sea breeze . . . Fresh air
and the golden rays of the fading sun fill the city's chambers."[2]
The ideal has hardly changed. Technology has advanced, and our
world has been transformed, but the preoccupation with urban
optimization and quality of life is timeless.

Future cities are traditionally the domain of urban planners,
architects, and social theorists, but today a new player is entering
the arena, using new tools to chase the same ideal: multinational
computing giants. Companies like IBM, Cisco, Siemens, HP,
and Microsoft are jockeying to build (and program) the city of
tomorrow, with the persistent goal of efficiency and well-being.
Just as Le Corbusier's unrealized Ville Radieuse was a reaction
to the contemporary automobile and the rise of mass production, today's smart cities are engineered as computer chips,
designed to address urgent considerations of sustainability and
efficiency.

With technology companies concerting their energies,
creating so-called smart cities seems imminently possible.
In addition to technological feasibility, the anticipation of
new markets has spurred investment. Construction of Songdo,
South Korea, began in 2004, with the heavy involvement of
Cisco Systems. The new city development is suffused with
digital technologies at every level.[3] Songdo was designed

Songdo Central Park

New urban-oriented partnerships between governments and software companies are rushing to build, inaugurate, and control the next generation of smart cities. With a tabula rasa planning principle, the hope is to integrate technology at every level, allowing unprecedented optimization and control. In Songdo, South Korea, buildings are designed for ecological sustainability, a smart transportation system monitors traffic patterns in real-time, and a pneumatic trash system pulls waste from individual homes to a central sorting and processing plant. This level of complete top-down control can be seen as stifling—if not Orwellian and dystopic. Do we need such full-scale experiments to inform smart city development? Pictured here is a view of Songdo's Central Park (with integrated saltwater-flow sensors, of course).

from square one as an integrated hardware and software system, capable of monitoring and controlling everything from transportation to utilities. Regardless of its success or failure, Songdo has attracted worldwide scrutiny and serves as a model for a host of subsequent projects.

The smart city presents the same appeal that planners have long dreamed of: when every element of a city is thought of coherently, the whole can function like clockwork. Rather than a "machine for living," what some now seek to create is an urban-scale circuit board, a computer in open air, driven by the objective of efficiency. Today's smart city is an engineer's or computer scientist's dream come true. Every piece of information is instantly revealed, and the urban machine can be controlled and optimized. With such a high degree of integration, information technology companies are seeking to enhance urban function: the race for urban optimization is in full tilt.

In short, cutting-edge cities are now shaped by the advent of pervasive information technology, enabling real-time connection, interaction, and communication. A similar transformation happened almost a century ago, when radio enabled the wireless transfer of information at the speed of light. A new era in communication had dawned. Even in the 1920s, it seemed that "when wireless is perfectly applied, the whole earth will be converted into a huge brain."[4] Collapsing space by instantaneously transferring information was set to revolutionize society.

In 1950, the mathematician and philosopher Norbert Wiener predicted a future in which "messages between man and machines, between machines and man, and between machine and machine, are destined to play an ever-increasing part" in society, concluding that "to live effectively is to live with adequate information."[5] Bringing that to the urban scale gives us the simplest definition of "smart city" as constant, ubiquitous high-volume information flow at the intersection of habitation and computing.

This information flow generally has three components. First, instrumentation: an omnipresent array of sensors measuring environmental conditions and movements (both human and material). Second, analytics: the algorithms that consume massive amounts of urban data to find patterns and even predict future scenarios. Third, actuators: digitally controlled devices that can respond to data in real time and impact physical space. These three main elements are united and governed by standard protocols for managing digital urban systems. As a whole, this software and hardware structure pushes forward the latest in sustainability technologies integrated at the building level—from occupancy-sensing architecture to resource-saving utilities.

This condition is commonly referred to as ubiquitous computing, or the "third wave in computing." Early mainframe computers, each shared by many different people, evolved into the personal computers we know today. And yet the screen-

and-keyboard-based interface is awkward. Mark Weiser, a Xerox PARC engineer who coined the term "ubiquitous computing," imagined that screens would disappear altogether as technology gently suffused cities, bridging digital and physical space. His was "a new way of thinking about computers in the world, one that takes into account the natural human environment and allows the computers themselves to vanish into the background."[6]

An environment of ubiquitous computing could also support a robust ecosystem of machine-to-machine communication across physical space. The evocative moniker "Internet of Things" (IoT) suggests that if individual objects could be imbued with a digital connective element, collectively they could become a physicalized network.[7] Anything and everything could be tagged and brought online. The refrigerator could touch base with the milk carton to make sure it was full and fresh and, if not, ping the nearest grocery store to stock a gallon of the same organic 2%. A world full of interconnected objects would create an unprecedented Internet-like structure in physical space. The ubiquitous geospatial mesh of IoT could have disruptive ramifications in all dimensions of business.

The trend toward ubiquitous computing is poised to transform the basic experience of urban habitation. As digital systems slip quietly into the background, an entirely new generation of consumer products will be introduced—"everyware"—imagined as intuitive, integrated, and invisible, an unobtrusive class of devices and systems that scarcely demand any attention from users. Everyware will become an ecosystem of quiet technology, deeply assimilated in urban space. Using that infrastructure, every element of the city and its buildings could be designed to derive maximum resource efficiency by working coherently and systematically.[8]

An "increasingly dense and intense information loop" based on digital systems has seamlessly merged with the age-old dream of urban optimization.[9] Cities are being galvanized by technology, bringing efficiency within reach. How Le Corbusier would smile if he could see an urban operating system optimize the smart city in real time!

But is ruthless, systematic optimization the most desirable application of ubiquitous computing? Is clockwork the only vision of "urban smartness" that should guide development? The iconic (and ironic) question remains: "How smart does your bed have to be, before you are afraid to go to sleep at night?"[10]

An altogether different application of ubiquitous computing and urban-scale digital networks has been emerging in parallel —a vision of a city based less on technology-driven optimization and more on the empowerment of people. Social platforms are connecting individuals and allowing communities to form around concepts and causes, both positive and negative. Most importantly, these shared ideas do not remain sequestered in virtual space. In 2011, sentiment that had fomented online spilled out onto the streets when cities across the United Kingdom were torn apart by the so-called Blackberry Riots. Platforms like Twitter spread catalytic phrases such as #TottenhamRiot or #LondonRiot, and over the course of five days, arson, looting, and clashes with law enforcement burned across the country. "Everyone watching these horrific actions will be struck by how they were organized via social media," Prime Minister David Cameron told an emergency House of Commons meeting. "We are working with the police, the intelligence services and industry to look at whether it would be right to stop people communicating via these websites and services when we know they are plotting violence, disorder and criminality."[11] The government did not know how to respond to the speed, scale, and fervor of ideas spreading on social media platforms.

Yet the most influential Tweet of the event—shared with over 40,900 Retweets—was a different call to action. In the wake of violence, #RiotCleanup suggested that citizens respond positively to the chaos around them. Almost instantaneously, #RiotCleanup gathered Twitter users in the same streets that had just been shattered by violence. The momentum carried across the United Kingdom, and ultimately more than ninety thousand people were involved with the effort. The (now expired) #RiotCleanup website stated that the movement "started not as an organization but an idea, an idea to change what was a downfall in society into something positive."[12]

The success of #RiotCleanup proves that the bottom-up, self-organizing energy of cohabiting humans is being expanded by networks and can, in turn, be reinfused into physical space. Digital platforms catalyze community, bringing people together to co-create and fix their city. But what if we had more tools—digital tools—to act on the city around us? What if the same mechanisms of smart urban optimization allowed people to take ownership of their city and make improvements that only residents could dream up?

If the feedback loop closes, specific local wisdom and actions may directly affect metropolitan policy. "Leak the knowledge of the neighborhood into codified systems—like a backward wikileak . . . Activate a citizenry," advocates sociologist Saskia Sassen.[13] This is the potent idea of hacking the city: opening up traditionally closed information systems and breaking the entrenched mentality of optimized urbanism. Ultimately, people can be empowered to take an active role in their environment. Open source technologies might be used to aggregate knowledge, skills, and ideas from a broad and heterogeneous citizenry and actually make tangible changes.

The seeds of a non-optimized city—one that embraces a touch of chaos and unexpected vitality—were repeatedly sown in the history of urban planning. Through writing, criticism, and activism, Jane Jacobs proposed the idea that cities should have their own texture, life, and alchemy, noting famously that cities are places for people. She advocated an interwoven planning concept, prescribing a set of specific components to generate diversity and heterogeneity. By commingling, many interacting elements would generate a vibrant and active urban space, resulting in such phenomena as "natural surveillance"—the idea that a stimulating urban fabric would draw people out and about at all hours and that their "eyes on the street" would ensure a safer environment for all.

Jacobs emerged as a champion of the citizens' city in the face of her contemporaries' uncompromising approach—most contentiously, Robert Moses' highway-based urban efficiency. Jacobs mounted what she herself called "an attack on current city planning and rebuilding," arguing that there is a higher goal for urban design than promoting high-volume traffic flow.[14]

The tension between these urban-planning mentalities, Moses versus Jacobs, essentially matches the divergence within architectural design, computer science, and politics known generally as top-down versus bottom-up. These two basic schemas define information flows, decision-making, and project direction models across several fields. A top-down approach first considers a concept at the broadest universal level and then systematically breaks it down into smaller and smaller components. Conversely, a bottom-up approach begins with the most atomized unit and builds it up into increasing complexity. In the case of urban planning, this is the difference between starting with an elegant highway scheme and cramming houses between its routes and starting with houses, shops, and parks and grouping them together into a city.

An alternate smart city could incorporate a bottom-up model—counterintuitively, it would not necessarily be at odds with data-driven urban systems. Network technologies allow for fine-grained control over physical space—the same control that can, of course, be used for mechanistic efficiency. But it may also become a tool for citizen engagement, actively involving the broader population in decision-making and operation. For example, a traffic system of autonomous cars could be optimized for maximum throughput, or for maximum sharing within social networks, or for maximum novelty and surprise. Given that a digitally integrated city could potentially be optimized according to myriad variables, the choice of which to focus on affects not only current function but also the subsequent array of possible choices. Planning decisions we make today determine the scope of choices we will have tomorrow.

Put simply, this is the concept of futurecraft in urban space. A merger of top-down and bottom-up systems can invite widespread engagement and mean effective implementation of solutions, ideally resulting in livable urban spaces. Pure optimization quickly becomes obsolete, but a hybrid model with a measure of chaos may be a more sustainable form of efficiency.

The idea of smart cities should be recast into something

more human-centric—what we often call "senseable cities." Optimization inflected with humanization means neither metropolitan-scale computers nor a network-enabled wild west. It is the convergence of bits and atoms: systems and citizens, interacting.[15]

In a dynamic and mutually reinforcing urban ecology, the historically fickle power of the crowd could be focused, while the slow-moving corporate and political juggernauts could be driven by the citizens they mean to serve. In the same way that free software—such as the computer operating system Linux—grew exponentially through open source development, engineered urban clockwork stands to benefit from sparks of serendipity. And just as ordinary people have hacked software, "citizen developers" can begin to hack their city. Various crowd-based platforms have proven the strength, ingenuity, bug fixing, and ideation of the world-at-large. A broad mix of experts, amateurs, corporate teams, and wildcard players is remarkably productive in unexpected ways if it can be effectively organized. Heterogeneity is beneficial. Yet space itself is crucial as well—the city constantly and definitionally provides new information to its agents, whether through challenges, collaborations, or inspirations. With the right frameworks in place, urban space can undergo an open source revolution similar to the one that transformed software.

Some examples have started to emerge. The New Urban

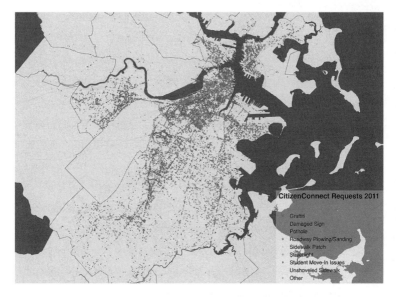

CitizenConnect Requests 2011

Graffiti
Damaged Sign
Pothole
Roadway Plowing/Sanding
Sidewalk Patch
Streetlight
Student Move-In Issues
Unshoveled Sidewalk
Other

Boston 311 Map by the MIT Senseable City Laboratory

We live in a hybrid digital-physical space, particularly in cities—but how can virtual networks, applications, and platforms tangibly change material urban reality? New tools are emerging that connect people and enable them to take an active role. One example, a new class of applications called 311 apps, provides citizens with a phone number to dial in and report urban issues, such as broken pavement, litter, or graffiti. The Senseable City Lab created this map of Boston's incident reports, sorting calls by type of damage and total number of reports. The data began with the start of the service in October 2010 and ran until June 2012, showing how citizens care for the city they live in over time. As more and more digital-physical platforms are integrated into urban spaces, the smart city may be designed, built, and operated in no small part by its "smart citizens."

Mechanics—an IT unit started in the office of Boston's mayor—
seeks to function as a nimble interface between citizens and
cities with the goal of promoting engagement and action. The
self-proclaimed "in-house R&D shop for the city" has grown
and taken root in different cities since its founding in 2010.[16]
Ushahidi, an entirely bottom-up organization, creates open software platforms for crowdsourcing information during disaster
situations, facilitating aid and enabling beneficial interventions.
Ushahidi started by collecting and mapping reports of violence
in the wake of Kenya's 2007 presidential election and has been
active in many disaster situations since its founding—from the
2010 earthquake in Haiti, to BP's Deepwater Horizon oil spill,
to Hurricane Sandy in New York—creating software platforms
that are instrumental in saving lives.

But is citizen hacking enough? Can bottom-up projects cause
deep change, transforming cities beyond visual gloss or niche
interventions? Short-term results must be immediately visible to
encourage ongoing participation—but what is the most effective first step? Does naming and nurturing an urban initiative
actually deter the most brilliant and iconoclastic would-be civic
hackers? What is necessary to spark bottom-up activity? "It will
take a lot more than civic-minded 'add-ons,'" writes the historian
Catherine Tumber, "to secure a culture of human agency—much
less a democratic one—in the hulking Corbusian mega-projects

fast profaning the landscape."[17] Allowing citizen participation requires vulnerability, slackened control, and the possibility of failure. But if hacking catches on, the productive integration of top-down and bottom-up urban paradigms may yet realize tomorrow's city by futurecraft.

PART II METROPOLITAN INFORMATION FLOWS

Not only have we already put the contents of the most important libraries of the world, and likewise the archives and museums and newspaper annals of every nation, on our punch cards, but also a great deal of documentation gathered ad hoc, person by person, place by place . . . What we are planning to build is a centralized archive of human kind.

Italo Calvino, 1968

FOUR
BIG (URBAN) DATA

In a short story, the Italian writer Italo Calvino imagined a society—a gently terminal dystopia—in which every detail and every moment is recorded for posterity.[1] All information would be compiled into the greatest document ever conceived, blending details of every individual life. The short story is problematized by intrigue and paradox surrounding information access, control, and deletion, and a final shocking twist puts an uncanny spotlight on the precipitous condition of absolute archiving. How will humanity remember itself? And how will it act when it knows that it is being recorded? These are prescient questions for a society that, today, is confronted with a similar situation of total recall.[2] Data are turning Calvino's fictional world into reality.

Consider every phone call, retail purchase, mile driven, mile run, Tweet, text, load of laundry that took place in the past twenty-four hours. Humans leave these kinds of virtual traces every second of every day, particularly in cities, and each is stored in a digital database. We are creating, archiving, and recalling digital copies of our world, forging a collective memory. "What would happen," Bill Gates once asked, "if we could instantly access all the information we were exposed to throughout our lives?"[3] It seems that, now, we can.

As digital technologies become increasingly pervasive (and networked online, as with the Internet of Things), every individual is generating a staggering amount of data, all of which are compounded across the entire population. Eric Schmidt, arguably one of the key individuals behind the big data revolution through his tenure at Google, noted that every two days humans create as much information as we did from the dawn of civilization up until 2003, approaching five exabytes of data (an exabyte is a quintillion—10^{18} bytes).[4] As digital information is captured and stored, the virtual copy of our world becomes increasingly high-resolution.

Any dataset collected for a specific purpose has an array of potential data by-products. Working with this deluge of urban information is what researchers often call opportunistic sensing: using data that have been generated for a specific reason and analyzing them in a different context to arrive at new conclusions. Datasets are often enriched with many dimensions, and whether or not every one of those dimensions is intended to have an explicit use when it is created, every aspect of the data can subsequently be instrumentalized in unexpected and creative ways. Credit card transaction data, for example, include unique IDs for the vendor and the consumer. These tags allow researchers to filter the data by location and type of purchase (food, gas, clothing) to understand patterns of economic behavior in cities.[5] Analyses of telecommunications data and social media have proven them both to be powerful tools for understanding human networks and dynamics.

Datasets can be considered individually, but far greater insights lie at their intersection and superposition. Geographic space is a common denominator that allows them to be linked. Particularly now that different streams of information can be interwoven and connected, urban data may offer an ever-clearer view of the human condition. As early as 2006, researchers began compounding telecommunications with transportation data.[6] The aggregate urban portrait that emerged—specifically during such extraordinary events as the final match of the 2006 soccer World Cup in Rome—revealed collective behavior tied directly to the event. Before the game, movement and communication were frenzied, activity slowed almost to a stop during the match, spiked sharply at halftime, fell almost to zero during the tense final minutes, and exploded when the match was called. Communication traces during the following hours revealed mass movement into the downtown area to celebrate the national team's victory. Subsequent projects in cities with more readily accessible data, such as Singapore, have compounded even more datasets. Data from weather, shipping, social media, public transit, cell networks, and more flow together to create a multidimensional portrait of cities and their patterns.[7]

In addition to opportunistic sensing, data can also be generated by deploying an array of sensors with a specific intent. Embedding technology into the urban environment can yield robust and fine-grained data, whether to map an existing system, to reveal dynamics that have never been brought to light, or to gain a new understanding of humanity's fingerprint. On a macro scale, the Google Street View car, for example, has driven across the world photographing 360-degree panoramas. After its first five years of operation, the Street View team announced that the car had captured five million miles of road in thirty-nine countries, generating a staggering twenty petabytes of data—quadrillions of images.

As more and more of these digital elements are embedded in physical space, many other aspects of the urban environment

Real Time Rome by the MIT Senseable City Laboratory

Data analysis and visualization can reveal important dimensions of a city and how its population lives—yet even more insight may lie in the merger of several related datasets. Real Time Rome was one of the world's first explorations of aggregated data from cell phones (through Telecom Italia's location platform), buses, and taxis to better understand urban dynamics and explore a data-rich future scenario through futurecraft. With maps such as this one, revealing the pulse of the city of Rome during the soccer World Cup (as Italy wins, playing against France in Rome!), the project shows how technology can help individuals make informed decisions about their environment, such as choice of transportation. The project points to possible strategies for reducing the inefficiencies of present-day urban systems through bottom-up behavioral change.

can be revealed—for example, the waste disposal system. As described in chapter 1, the Senseable City Lab began a project, Trash Track, that addressed the scenario of ubiquitous tracking. Researchers created geolocating tags and worked with residents of Seattle to attach them to thousands of ordinary pieces of garbage—effectively creating an "Internet of trash"—to map the waste removal chain across the United States.[8] Over the following months, the devices revealed a surprising network that had been completely unknown before. In the future, an accelerating diffusion of technology into urban space may offer an unprecedented understanding of systems like waste management dynamics and may create data that can be used to optimize the entire system, even in real time.

The trend points toward a phenomenon that has been termed "smart dust." Physical space could be laced ubiquitously with nanosensors—scattered micro devices that are smaller than grains of rice. "Large-scale networks of wireless sensors are becoming an active topic of research. Advances in hardware technology and engineering design have led to dramatic reductions in size, power consumption and cost . . . This has enabled very compact, autonomous and mobile nodes, each containing one or more sensors, computation and communication capabilities, and a power supply."[9]

A rich array of data will be available in a future scenario of ubiquitous smart dust. In the meanwhile, a pervasive network

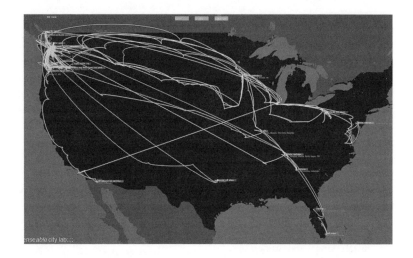

Trash Track by the MIT Senseable City Laboratory

Trash Track, a project by the Senseable City Lab, is an exercise in futurecraft. Researchers imagined a future in which almost everything could be tagged and tracked, and information was constantly streamed online. Trash Track was first deployed in 2009 in Seattle, Washington, bringing together hundreds of volunteers and thousands of geolocating tags. Citizens attached GPS tags to garbage and disposed of it as usual. Digital and physical merged to reveal an invisible metabolic function of the city: its waste removal system. Over the following weeks and months, the sensors traced a dizzyingly complex disposal chain across the entire United States. An elaborate national waste removal system was brought to light through technology that disappears into the environment.

tem, one that can sense the world, create data, and respond. Pictured here is a microelectromechanical system (MEMS) with a tiny mite perched on top to show its size. Researchers are currently working on refining a wide array of capacities for such devices, including inertial sensors, microfluidic actuators, photovoltaics, and advanced optics. With the continued development and deployment of this "smart dust," digitally augmented space will open new avenues for research and urban applications—and possibly a future in which using a computer does not involve looking at a screen while manipulating a mouse and a keyboard.

Spider Mite on Mirror Assembly by Sandia National Laboratories

Digital technology is shrinking, approaching a scale at which it can disappear into the environment. This suffusion enables what is known as a cyber-physical sys-

already exists in our cities today: citizens themselves. In some cases, collecting personal data is fully intentional. The computer scientist Gordon Bell was one of the first to explore the idea of individual data in a practical way—in 1998 he began a project called Your Life, Uploaded, making himself the subject of the first full-resolution experiment in so-called life-logging. Bell created the hardware and software to capture every moment and every action of his life through photos, computer activity, biometrics, and more. The technology was primitive, and in many ways disruptive, but the project was successful in cataloguing his existence for more than a decade. "The result?" he wrote. "An amazing enhancement of human experience from health and education to productivity and just reminiscing about good times. And then, when you are gone, your memories, your life will still be accessible for your grandchildren."[10]

What Bell initially set out to do as a full-scale scientific and sociological study is now the unconscious norm—the default condition—of an Internet generation. Our spatial and social activity is tracked and logged (in many cases, it requires more effort and determination to opt out of documentation than to engage in it). Tweets, Uber calls, text messaging, Yelp reviews, and check-ins are becoming the natural activities of daily living. A trove of information is flowing out of and through the population, and people are more and more interconnected. The bulk of the extraordinary (and growing) amount of data being created today is user-generated content, an almost constant stream of personal data.

There are many platforms for users to upload and share photos, for example, and underneath their straightforward functionality lies an enormous and rich dataset, including the GPS location, keywords, time, social networks, and popularity associated with each image. This trove can provide a deep understanding of how people interact with and in physical space as digital traces are mapped and overlaid—revealing, for example, the movement and activity of tourists. Using Flickr data, Senseable City Lab researchers mapped a crowd-generated cartography of Spain, showing how visitors and residents see and use their environment, identifying, among other things, hotspots, or "visual magnets." [11] Researchers could effectively borrow the eyes of the population in a continuing analysis, applying computer vision image-processing and color-matching to landscape photos. The user-generated photographic data began to reveal natural ecological conditions like drought and urban green spaces. This futurecraft scenario extrapolates from the perpetually expanding global knowledge base, the fabric of a digital blanket that covers and suffuses the cities we live in.

Anyone and everyone can put tiny chunks of data online—and we do, almost constantly, whether intentionally or unintentionally. Some incentive-based platforms give individuals a small monetary compensation for completing a single task that contributes to a greater collective effort, while other systems enable deliberate civic action through a more altruistic motivation. Numerous community-based smartphone applications automatically upload detailed information about road damage, traffic, and gas prices for the benefit of all drivers. Citizen contributors to the collaborative OpenStreetMap (particularly relevant in areas not yet graced by Google) draw roads and make the information publicly available. Many 311 apps allow city dwellers to report nonemergencies, including potholes, fallen trees, and

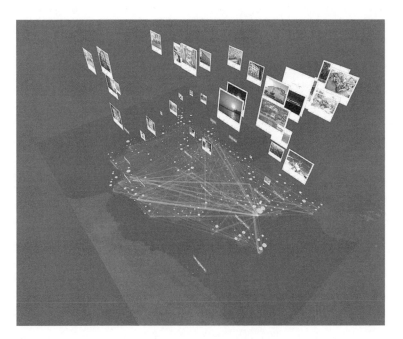

Ojos del Mundo by the MIT Senseable City Laboratory

Almost every human alive today is somehow represented online: each of us creates data, constantly and unconsciously. Some of this information we share willingly—for example, by using social media platforms. The resulting big (human) data can show how we interact with our environment and with each other. Using data from the photo-sharing platform Flickr, researchers at the Senseable City Lab were able to see the world through the eyes of its inhabitants, whether tourists or residents. The project, titled "Ojos del Mundo" in Spanish, or "Eyes of the World," mapped social photo-to-sharing activity across Spain to identify movement patterns, attractions, and even—by applying color analysis to photos—areas of drought.

damaged street signs, and either alert city government or organize the community to fix the problem. Another category of data is unintentionally created, encompassing the broad (and expanding) array of social media platforms, such as Twitter, Facebook, and Flickr. Users may hardly realize that their actions online are a rich source of data—a fact that was exploited for a research collaboration between Facebook and Cornell University, to much controversy.[12]

The three categories of data collection in cities—opportunistic sensing, ad hoc sensor deployment, and crowdsensing—can be hybridized to a various extent. On the backbone of telecommunications networks, a new universe of urban apps has appeared, allowing people to broadcast and exchange geolocated information and reveal the city from their personal perspective. Air quality, for example, is poorly understood because data are collected in static and sparse ground-based stations. In a possible future, citizens themselves could carry a distributed network of sensors that create a dynamic real-time atmospheric map. Using smartphone-integrated sensing devices, pedestrian commuters could generate data at the human scale, as though a tracer were running through the veins of the cities, showing the urban environment that the commuters live in and move through.[13] This concept may inspire manufacturers of consumer electronics to include environmental sensors and to publicly release the resulting data for analysis.

Individuals are becoming agents of data collection, and at the scale of an urban population, we all constitute a vast trove of crowdsourced information. As Gordon Bell's experiment recedes into the past, society is moving from "life-logging" to "city-logging." We are all enmeshed in a distributed sensing ecosystem.

Calvino imagined a situation of total recall, where every detail was recorded. What he did not imagine was that individuals would share that information willingly. This is a radical shift from a top-down vision of data collection to a bottom-up vision of data aggregation and dissemination—it is a shift from "Big Brother" to "little sisters." A (re)distribution of control over information may empower individuals, providing insight into the kind and quantity of data that they create and a choice of what and when to share—even, perhaps, prompting claims to some of the information's value. Today, for example, Gmail users let Google read their e-mail and send them targeted advertisements in exchange for using the service free of charge.[14] Researchers have proposed that, in the future, a personal data management tool—a "data box"—could explicitly give individuals the choice to keep data private or to upload it freely and receive benefits in return. This idea is effectively a "mutation" of our present system and might emerge from an active discussion among all involved parties. The most desirable future might be one in which people have the opportunity to directly benefit from the inherent value of their daily actions.

Broad trends point toward an increasing number of apps and technologies that create urban data, wider adoption across demographics, and faster response from city governments. Entirely new positions are being created at the city level—urban CTOs (chief technology officers), for example, will nimbly manage the urban implications of digital systems on the macro scale, while personal data management tools will govern on the micro scale. People are quantifying themselves to better understand who they are and how they can improve their lives. Together, we compose a mosaic portrait of the city. People are becoming more and more aware of the digital shadows they cast and will be empowered to take a more active role in inhabiting—and contributing to—the places where they live. We are moving from the quantified *self* to the quantified *city*.

*You are a cyborg every time you
look at a computer screen or use a
cellphone device.*

Amber Case, 2010

**FIVE
CYBORG
SOCIETY**

Cyborgs, creatures of intertwined biology and technology, have fascinated humans for decades, at least since the term was coined in the 1960s. One of the first cyborgs was a rat equipped with an osmotic pump; the enmeshed technology was intended to function as life support in hostile environments. The project pointed toward the broader goal of "adapting man's body to any environment he may choose," or, more specifically, creating a human-machine system that could one day maintain homeostasis in outer space. The key to unlocking new environments was to reconsider the human body itself: researchers sought to "permit man's existence in environments which differ radically from those provided by nature as we know it."[1]

A cyborg is a hybrid capable of more than either the biological or the mechanical system alone can do, with a correspondingly expanded range of possible habitats.

Given the general definition—dependence on synthetic tools for inhabiting new ecosystems—cyborgs are not strictly science fiction. In a basic sense, the human is a cyborg species, distinguished (though not unique) for its creation and appropriation of extrinsic tools. Weapons and fire and clothing, for example, enabled our ancestors to inhabit environments they otherwise could not, just as the osmotic pump would serve the rat bound for outer space. Humans create technologies that surround the body and support its physical survival.

Human progress, on a macro scale, can be tracked through the use of tools, from the earliest worked stones in the Paleolithic Era to mechanical technology in the Modern Era. Life-support technologies have become increasingly sophisticated over the course of human history. During the 1960s, new theories sought to articulate the cyborg condition, finding humans to be nothing without a constellation of external tools. The prevailing idea was that, rather than using tools as accessories for augmenting specific functions, humans do not exist independently of a collage of support systems. "The late 1960s saw humanist disciplines contend with new human sciences that revealed 'man' to be a construct by exposing his technologies as supports and extensions."[2] The French anthropologist

André Leroi-Gourhan mapped this transition in his touchstone work, *Le Geste et la Parole,* drawing a curve of human progress across history according to the use of tools, from the Neolithic Era to the twentieth century, from primitive stone utensils to the development of digital technologies.[3]

Following the progression from stone to silicon tools, we could describe humanity today as a new form of *Homo sapiens—* one with an entirely new apparatus. Tools have historically been a modification of the physical self, extending such capacities as speed and strength; essentially, they are implements that transform matter around us in ways that are difficult or impossible using just the human body. Yet today's tools are fundamentally different, extensions not of the anatomy but of the mind—of memory, identity, and social function. "Human progress was marked by the gradual externalization of functions," wrote Antoine Picon, "from stone knives and axes that extended the capacity of the hand to the externalization of mental functions with the computer."[4] The erosion of the discrete boundary around a human being occurred at this distinct shift in the function tools.

Discourse evolved as technology played an increasingly central role in daily life—approaching prosthesis. In the 1980s a nascent "cyborg theory" emerged, positing the cyborg condition as a new paradigm of human social-biological-technological existence. The feminist theorist Donna Haraway articulated a social dimension of cyborg theory and propelled the idea into broader public debate. Humans, she wrote, are "post–Second World War hybrid entities made of, first, ourselves and other organic creatures in our unchosen 'high-technological' guise as information systems, texts, and ergonomically controlled laboring, desiring, and reproducing systems. The second essential ingredient in cyborgs is machines in their guise, also, as communications systems, texts, and self-acting, ergonomically designed apparatus." And this is the crucial feature of the modern cyborg: digital technologies have become a dynamic extension of our bodies and minds, demanding a constant and two-way cybernetic exchange in a way that our traditional (one-way) extensions, such as clothing or axes, have never done. This has even been conceptualized as a "third wave" of civilization, following radical transformations in mobility and telecommunications technology—what the urban cultural theorist Paul Virilio calls a "revolution of transplants."[5]

Virilio further contends that human existence is "no longer structured by the polarity of public and private."[6] Humans are thrown, exposed, into the world, but we are also sustained by the world. In many ways, Haraway's cyborg is the citizen of McLuhan's convoluted, connected, and often contentious global village. For the human as sender and receiver of information, a modern social context forges and maintains entity indistinguishable from identity.

The term "posthuman" describes a new entity that is born with technology rather than acquiring it as a prosthetic. The broad (and expanding) spectrum of critical discourse surrounding our posthuman future seems to vacillate between the threat of technology's corrosive force on humanity and the wildly optimistic eulogy of biotechnical augmentation.[7] The common denominator is a deep entanglement of human and technological systems. "We of the modern age are provided with two types of bodies . . . the real body which is linked with the real world by means of fluids running inside, and the virtual body linked with the world by the flow of electrons."[8] Those two bodies are, today, inextricably enmeshed and coevolutionary. One piece of technology forges the strongest link, arguably more transformative than any other: the smartphone.

Smartphones are effectively powerful mini-computers enhancing humans' logical and computational capacities, particularly because they are always available. Take, for example, remembering an address: rather than committing it to memory, that information can be stored in a digital contact book or quickly looked up online at any moment. Memory has been outsourced. Furthermore, we have abdicated any navigational responsibility for reaching a location in favor of individualized, personalized, on-demand way-finding tools such as Google Maps. The tools for mapping can respond to real-time traffic information and then deliver suggestions for a good meal on

the way. The theorist Bruno Latour, together with Valérie November and Eduardo Camacho-Hübner, explores this reconfigu-
ration of maps as dynamic "navigational platforms" in the 2010
paper "Entering Risky Territory: Space in the Age of Digital
Navigation."[9]

The posthuman is a creature born into this binary condition,
into a world of converged digital and material, where each
individual's mental and social existence is enabled, sustained,
and improved by technologies. Beyond individual personal
interactions, the global adoption of smartphones—mass mobile
communications—amounts to a collective societal shift. Over half
of the global population is now instantaneously interconnected.
Personal devices serve as a portal to externalize and multiply the
self to a conceivably infinite degree. The prosthetic smartphone
has deeply permeated society along the backbone of wireless telecommunications, giving rise to a new networked humanism.

Anthropologists have considered this aspatial and omnipresent social phenomenon. Connective technologies generate a
condition of "ambient intimacy," wherein people have access to
each other—whether in real time or as digital avatars—at any
moment they choose. As this becomes pervasive and integral,
humans in the networked society will be co-creating each other
at all times.[10] Tools have evolved from physical to mental to
social.

But the story does not end here. Technologies have made a

radical pivot back to physical space.[11] Digitally networked humans are at the crux of bits and atoms: integrated technologies are profoundly transforming not only social identity but corporeal inhabitation of cities. "The figure of the cyborg is at root a spatial metaphor," notes the geographer and urbanist Matthew Gandy. "But how does the idea of the cyborg intersect with spatial theory? In what ways does the cyborg reinforce or contradict other emerging strands of urban thought that also emphasize urban complexity and hybridity?"[12]

Through smartphones, the daily lives and modes of perception for an Internet generation have become a posthuman condition, with architectural, spatial, and—in particular—urban implications. "This generation has developed physical and mental attitudes that call for a different kind of space, a space that can be deciphered through systems of clues and series of unfolding scenarios."[13] Through smartphones, the city is now burgeoning and constantly unfolding inside every pocket. Every citizen has a tool with which to perceive and process the city.

Peering through this digital lens is an intensely personal experience. Your smartphone locates you precisely in space and time, and it knows your preferences, schedule, and consumer patterns. Geolocated applications "no longer adhere to the anything-anytime-anywhere-new-media paradigm of the 1990s. Rather, they are centered on location-sensing capacities and aim to intervene in or add to a specific here-and-now. Their exact interventions differ, but . . . urban media are making deep inroads on a diverse range of activities of place making—be they the top-down deployment by government agencies or the bottom-up appropriation by urbanites in their every-day life."[14]

This class of smartphone applications, known as location-based services (LBS), was introduced around the year 2000, offering a menu of personally relevant information through location sensing. LBS entered a variety of domains, from mobility, entertainment, and food to way finding, weather, and romance—all enabled by the ubiquity of smartphone devices. One of the earliest, called Dodgeball (launched in 2000), allowed users to log in with their location and receive notifications about crushes, friends, friends of friends, and interesting venues in their general vicinity through SMS text messaging. Google acquired Dodgeball in 2005 but discontinued the service in 2009; according to critics, "It didn't make the company any money, and its user base had shrunk to a small cadre of digital-media enthusiasts based primarily in New York. It's sort of the Arrested Development of Web apps: not particularly popular, and most people don't even seem to really understand it, but those loyalists sure are loyal. And much like a TV show with a small fan base, the corporate parent pulled the plug."[15]

Yet shortly after Dodgeball's demise, its original founder, Dennis Crowley, launched a new project called Foursquare. Unlike his initial foray into LBS, Foursquare became wildly successful.

Foursquare is different from Dodgeball in key ways: it is an app unto itself, it offers incentives ("You're the mayor of ———!), it integrates a "To Do List" and "Tips"—but most importantly, it emerged at the right time. By 2009, smartphones had become nearly ubiquitous, bringing with them a new dimension of the city and a new mode of interacting with and in urban space. "The abstract motifs that compose this landscape confer on it something mysterious, even magic, a re-enchantment of the world."[16] This re-enchantment can (and does) happen most potently in cities. While smartphones certainly exist in rural areas, urban agglomerations offer the density and critical mass that can make applications like Foursquare successful.

A common criticism of digital applications—specifically, one leveled by the voices predicting the "death of distance"—was that they would eliminate the chemistry of the city. Yet real-time applications are now delivering rich, digitally brewed serendipity, and rather than neutering urban space, networked systems are becoming a new interface with the physical world. Each smartphone communicates in real time with a constellation of phones, businesses, and networks surrounding it.

Systems are becoming increasingly robust, enabling real-time everything, from Uber (a platform for citizen taxi cabs) to Tinder and Grindr (connecting nearby people for anything from social to romantic encounters). Always-on devices connect the majority of the human population to one another, to physical places, and to dynamic processes. Cyborg humans have entirely new modes of inhabiting the physical city at all times.

As these networks become increasingly dense, the smartphone remains an integral tool for data generating, interface, and collision, for combining and contributing to various streams of information. Local and personal information delivered through a smartphone is a mediated, bite-size connection to global networks and the larger ebbs and flows of metropolitan function. Through your smartphone, you can understand and digest the broader, complex reality of the city. It serves as a control room, revealing urban systems such as transportation, weather, and social and interactive media. Understanding these urban dynamics enables people to more effectively (or enjoyably) inhabit the city. The smartphone is a location-inflected analytics tool with almost infinite dimensions.

Smartphones, and their constellation of associated systems, are not without behavioral ramifications. Data, networked platforms, and connectivity can provoke the kinds of collective action that transform cyborg citizens into a smart mob. Through location-based media, there is a broad emergence of swarming behavior, a new social phenomenon wherein groups of any number are digitally interconnected for any reason and begin to act en masse. Amorphous and porous smart mobs do not need to have a specific agenda, nor do members need to have any prior familiarity with each other.[17] They are flexible groups of people who have been connected by digital media and converge in physical space.

As online and on-site activity blend together, crowds can be cohesive and, potentially, productive. Two simultaneous trends—increases in both smartphone adoption and embedded technologies—are transforming what was formerly a communications network into a sensing network. A whole class of applications appropriate embedded hardware in the smartphone that is intended for other purposes—for example, using a phone's accelerometer to detect pace, or its camera light to measure heartrate. These take advantage of the always-on prosthetic device to implement "viral sensing" at a large scale.

In addition to these opportunistic apps, another class of digital technologies links the phone to external hardware that extends its capabilities. These piggyback technologies work symbiotically with the phone—taking advantage of its high-power computing, high-speed network access, and nearly constant use—and have given rise to the quantified-self phenomenon. A variety of quantified-self gadgets, from bracelets to pins to watches, can tell users everything about their daily activities, including steps taken and patterns of sleep. Not only are smartphones our mechanism for interfacing with the world around us, augmenting our social and professional selves, they are now a means of scrutinizing our individual bodies, our biological selves.

The smartphone, with its constellation of associated hardware, is a first step toward the seamless exchange of information to and from the human body: data are constantly recorded, uploaded, and downloaded in real time. The first generation of truly commonplace cyborg technologies includes pacemakers (mechanically supported hearts), cochlear implants (allowing the deaf to perceive sound), and visual prosthetics (cameras that input directly to the brain). In *Rebuilt: How Becoming Part Computer Made Me More Human,* the technology theorist Michael Chorost characterizes the activation of his cochlear implant as a transition to becoming a cyborg.[18] Yet these devices are all one-way transfers of information. The next step is quickly approaching as a variety of wearable and ambient devices become constant two-way conduits of information between the body and the network. Just as the Internet of Things is connecting billions of devices worldwide, digitally integrated implants are creating a new interface with our physical anatomy. In addition to providing a machine-to-human interface, these implants may also enable machine-to-machine connection and analytics. Humans are becoming directly enmeshed with the network.

As technological sophistication advances, could devices begin to sense the human biome—even map humanity's evolving health? Will we soon be communicating, geolocating, and tracking nutrition through a "digital tooth" implant? Will our contact lenses—and then our irises—become both cameras and real-time augmented reality displays? And what are the privacy issues? Google Glass—a project by the tech giant—was progressing in this direction, integrating glasses with a constant heads-up, voice-controlled display. However, after a negative public reaction and weak consumer response, and amidst an explosion of press, the project was put on hold before it was ever released.

More and more bodies are going online, and detailed external analysis of the collective quantified self is becoming possible. At the moment, quantified-self technologies do not go far beyond confirming what you already know. When you wake up in the morning, your bracelet-integrated phone will announce that you slept poorly—the last thing you want to see as you rub your eyes and blink through a headache. But in the future, this kind of data will extend beyond the individual and reveal broader human patterns—whether through the work of specialists or by harnessing the collective intelligence of the crowd. A new "quantified us" paradigm might map the human biome on the community, city, or country scale.

In light of the collective quantified us, the idea of a singular cyborg is being recast as an Internet of Bodies. Through networked technology, cyborgs—cybernetic organisms—may become a cybernetic species. There have been dire predictions of two scenarios: that computers become conscious or that

consciousness is uploaded into a computer. The apocalyptic concept of the Singularity is that "we will soon create intelligences
greater than our own." When this happens, when machines
become an autonomous, aware, cohesive (and malignant) system, "human history will have reached a kind of singularity."[19]
Yet what seems to be far more likely is that human and machine
will become seamlessly merged. In a way, this negates the fear
of an impending Singularity. Ultimately, the human-augmented
machine—that is, the machine-augmented human—will always
be superior to systems that are exclusively machine or human.
We have nothing to fear from machines as they become sentient
and symbiotically interlaced with human consciousness. The new
cyborg could be a networked human machine, (re)empowered
to enhance and augment the individual through contact with
others.

If I was to realize new build-
ings I should have to have new
technique. I should have to design
buildings that they would not
only be appropriate to materials
but design them so the machine
that would have to make them
could make them surpassingly
well.

Frank Lloyd Wright, 1932

SIX
LIVING
ARCHITECTURE

Sudden transformations sparked by abrupt technological leaps have punctuated the history of architecture.
During the mid-1400s, into the context of a craft-based
architectural tradition, Leon Battista Alberti introduced a
mathematical approach to graphic representation. In so
doing, he paved the way for Renaissance classicism: architecture focused on precision and representation through
drafting rather than approximate construction. Four centuries later, steel and glass enabled engineers like Isambard
Kingdom Brunel, Sir Joseph Paxton, and Gustav Eiffel
to design daring and innovative structures and shatter
the limits of what could be constructed. Soaring feats of

technological prowess became a new aesthetic at the nexus of architecture and engineering. A generation later, at the crest of the mechanical era, Le Corbusier appropriated the tools and forms of mass production, concluding—as previously cited—that the house is a *machine for living in*. Architecture was optimized not only from the standpoints of design and structural engineering but also from the standpoints of mass production and social function.

Technological upheavals are the lurching steps of architectural progress, its driving force. Le Corbusier dreamt of "realiz[ing], harmonically, the city that is an expression of our machinist civilization."[1] Yet our civilization today has transitioned from mechanization to computation. The digital revolution—the convergence of bits and atoms—is poised to be the most radically disruptive change that has ever recast the design, construction, and operation of our built environment. Just as machines brought standardization and high output, digital tools can bring dynamism, variation, and responsiveness. The question now becomes, How will architecture evolve in the digital era?

Initial attempts to address this question—to create dynamic architecture for the digital age—have been form-based. Designers have created evocative architectural sculptures that shout distinctive visual identities: Frank Gehry's iconic Guggenheim Museum Bilbao, for example, and the similar projects he has scattered around the world. These have ushered in a new

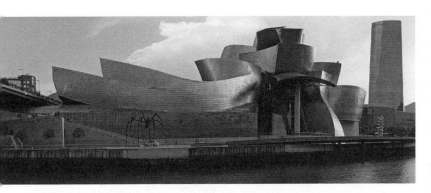

Guggenheim Museum Bilbao by Frank Gehry

For hundreds of years, painstakingly hand-drawn plans have defined the profession of architecture. As a result, buildings have been limited to what can be conceived and represented in a manual way. Today, powerful computation has brought a radical transformation. Using parametric design software, designers have pushed the boundaries of formal possibility, an experimentation driven by the goal of creating vibrant, dynamic, and "living" structures. The architect Frank Gehry, one of the defining voices of his generation, has developed a signature style, often using metal panels to create a skin for complex curving shapes. Pictured is the Guggenheim Museum Bilbao, a contemporary art museum inaugurated in 1997 and arguably Gehry's most widely recognized and praised project.

aesthetic regime of irregular and organic buildings, often called "blobby" architecture.

The new formal language was enabled in large part by parametric design software: digital tools that allow the architect to script an internal logic, input data values (objective contextual factors, zoning, or functionality requirements), and run an algorithm to negotiate those constraints and produce formal, often extraordinarily complex artifacts. Rather than detailing intricate specificities by hand, the architect writes parameters, and the computer churns out highly elaborate results.

Energetic swooping shapes and structures became possible at the architectural scale. Parametric software opened a new arena where designers could radically question inherited formal assumptions about architecture. They explored the boundaries of possibility eagerly and productively, assuming that—given an opposition between rational and organic—nongridded and complex forms have a more vibrant quality. Early theorists of parametric architecture characterized a new sensibility that aimed for "maximal emphasis on conspicuous differentiation."[2]

The highly visible 2004 Venice Biennale of Architecture, titled *Metamorph,* explored the "fundamental changes under way in contemporary architecture, both in the theoretical and practical design field, and in the use of new building technologies." The event brought together architects, academics, researchers, and critics at the forefront of computational design. Individualism and experimentation defined the collective rhetoric, but a more cynical view of the menagerie of projects found the differentiation to be superficial. "The computer has finally made possible forms that are different, at the same cost as the standard forms of old. A newness of very similar forms though, more sculptural than radical, buildings and structures with sensual folded, twisted and curving surfaces. It looks more like an international computer art festival . . . and the most important theme to come out of the biennale was the question of redundancy."[3] Under the guise of novelty, the common denominator that emerged was predictable manipulations of complex geometry rather than meaningful dynamism.

Parametric tools have granted architects an unprecedented power to generate space using algorithmic functions and to appropriate a rhetoric of vibrancy. As the trend has developed since *Metamorph,* however, architects have been hard pressed to find meaningful data to feed into algorithmic design processes. A cruise ship terminal in Japan, for example, was informed by the geometry of waves in traditional paintings, specifically "the Hokusai Wave." The designers were inspired by "a drawing from a local painter that we had been toying with while we indulged in geometric manipulations and construction hypotheses during the design phase."[4] Whereas this project was successful, the application of parametric software, in many cases, goes no deeper than the skin of a building. Algorithms can compute thousands of unique elements to compose a dazzling facade on an otherwise standard structure.

Parametric design promises a certain novelty, whether it is driven by geospatial data or by complex matrices of associations. The virtual dimension that now blankets physical space is burgeoning with data, some of them appropriated by designers to plug into scripts as they seek "to grow or evolve new formal configurations in response to specific forces and constraints: structural, climatic, or programmatic. While this has produced compelling formal results, there are conceptual and procedural limits. The design techniques used to generate these new buildings may be dynamic, but the buildings themselves are static." Architects can generate an almost infinite number of formal solutions in a given situation, but complexity and magnitude are not inherently meaningful or *living*. "The forms generated may resemble nature, but they retain little of the performative or adaptive complexity of life itself."[5]

Algorithmically generated architecture is a static visualization of larger complexities. To evoke the fluidity of digital space in an inert physical object is to freeze a dynamic process, as if pressing Pause to find a single frame in an action sequence. Even the climax of energy and vibrancy, caught in a still frame, will convey only a shadow of the dynamic whole.

Visual complexity can be computed, but can it deliver anything more than curb appeal? And is that even desirable? The digital age has already suffused our world with innumerable flows and layers and intricacies, and formal plasticity only adds visual chaos to the ambient complexity.[6] Digital tools could be integrated with architecture, beyond veneer or gloss.

How, then, to integrate digital systems to achieve true dynamism? "Being digital is not primarily about using a computer in the design process, nor about making this use visually conspic-uous. It is an everyday state that goes hand in hand with ges-tures as simple as being called on a cell phone or listening to an mp3 player."[7] That is, architecture should become an integral and responsive part of human life. Architecture must do more than just *look* like a living organism: it should *perform* as a living system.

The earliest glimmers of this possibility date back to ex-perimentation with moveable structures in the mid-twentieth century. A group of young Japanese designers, the Metabolists, imagined living architecture for the growing population of postwar Japan. Buildings, they proposed, could be shaped over time by the pushes and pulls of sociodynamic forces. Metabolist structures used biological models, attempting dyna-mism through, for example, spine-and-branch arrangements or cellularly subdivided megaforms. The architect would establish a master program (or "DNA") that could propagate itself according to a patterned structural system.

Few of their structures were ever built. One notable exception —Kisho Kurokawa's Nakagin Capsule Tower, located in central Tokyo—is a paradigmatic example of Metabolist theory. It is con-ceived as a central spine, onto which individual housing pods can be attached and rearranged. In theory, infinite combinations of pods and connections between them allow residents to create larger or smaller spaces in response to different families, budgets, or changes in housing demand over time. Yet the Capsule Tower reveals a deep conceptual flaw: since the building's completion in 1972, not a single pod has been shifted or combined.

The twentieth century is dotted with a handful of attempts to construct functional mutable architecture, but in most cases, the structures have fallen into stasis or remain unbuilt. An entirely flexible structure still requires inspired occupants to take agency. In practice, mutable buildings go largely unchanged.

Flexible structures may not spark active participation, but it is here that digital technologies reenter the playing field, enabling a more gentle, intuitive, and responsive interaction between humans and the built environment. Far outside the discipline of architecture, pioneering computer scientists and mathematicians of the mid-twentieth century started developing a theory of cybernetics. The emergent discipline sought to explore networks, focusing on communication and connections between interdependent actors in a system. Cybernetics, according to Gordon Pask, the academic responsible for popularizing the idea among architects, is "how systems regulate themselves, reproduce themselves, evolve and learn. Its high spot is the question of how they organize themselves." This conceptual framework could be productively applied to architecture. As a practical design strategy, cybernetics is about negotiating a set of interrelated factors such that they function as a dynamic system. "The design goal is nearly always underspecified and the 'controller' is no longer the authoritarian apparatus which this purely technical name commonly brings to mind. In contrast the controller is an odd mixture of catalyst, crutch, memory and arbiter. These, I believe . . . are the qualities [the designer] should embed in the systems (control systems) which he designs."[8] The architect becomes a choreographer of dynamic and adaptive forces rather than scripting outcomes in a deterministic way.

Around the same time, architects at the fringe of the discipline took the idea of interactivity and sensationalized it. Architecture became loud, fun, hip, and constantly evolving. Buildings were thought of as venues for action and interaction, as dynamic scenes that could incite events and connections and evoke delight. The Generator Project, by the architect-provocateur Cedric Price, was a clear exemplar of this new attitude. An unbuilt concept for a retreat and activity center, the project consisted of a system of 150 prefabricated cubes, each twelve feet on a side, that could be shifted and reconfigured—much like the pods in the Nakagin Capsule Tower—but, crucially, would also interact in a dynamic way. A primitive digital software detected inactivity, and if the building remained static for too long, the software automatically executed "The Boredom Program" to reconfigure its own structure and incite (or perturb) users. The architecture itself took an active role as provocateur, with the aim of enhancing human experience. This was a system for dialogue and mutual reaction, beyond the Metabolists' linear user-changes-building idea. In many ways, this work was an application of cybernetic ideas to the field of architecture: it created systems that would dynamically self-organize in response to inputs and actions.

If the first industrial revolution was concerned with creating machines optimized for a specific task, cybernetics, in contrast, was concerned with a new kind of (perhaps nonmechanical) "machine" that could satisfy an evolving program. "We are concerned with brain-like artifacts, with evolution, growth and development; with the process of thinking and getting to know about the world. Wearing the hat of applied science, we aim to create . . . the instruments of a new industrial revolution—control mechanisms that lay their own plans."[9] Translated into architecture, cybernetics means buildings that function as adaptive learning entities living in a kind of dialogue with their inhabitants.

Active and networked architecture is starkly opposed to recent form-focused attempts at dynamism, and may illuminate an alternative path forward. "Today, many designers have turned several late twentieth-century infatuations on their heads, for instance with speed, dematerialization, miniaturization, and a romantic and exaggerated formal expression of complexity. After all, there is a limit beyond which . . . complexity simply becomes too overwhelming."[10] Rather than using digital tools to mathematically calculate complexity for the visual sense, interactive spaces can use digital tools to generate a new form of complexity: experiential complexity. A shift away from elaborate structures and toward architectural dynamics entails buildings that perform as (rather than appear to be) living organisms.

Computation will not be used only to define intricate shapes according to parameters but will also become an integral part of the building, interacting with users according to a program. This interface functionality points to embedded rather than generative technology. In addition to plans and sections, architects in the future will be free to specify a system of interrelated sensors, operations, and actions—loops that bring architecture to life based on a dynamic set of experiential and functional requirements. Grounded in communication and learning systems, sensor networks can transform buildings into intelligent agents with the capacity to learn from and coexist with their occupants. The dream of dynamic spaces can finally be fulfilled as buildings weave together humans, environment, infrastructure, and personal devices.

Just as mobility networks are taking advantage of ubiquitous sensors (as with crowdsourced maps or pothole detection), so too will buildings take advantage of the human flows running through them. We will shift from living *in* a home to living *with* a home. Architecture becomes a form of interface, playing an active role in the human environment, both digital and physical. "The goal is to facilitate as seamless a movement as possible from fast to slow, virtual to physical, cerebral to sensual, automatic to manual, dynamic to static, mass to niche, global to local, organic to inorganic, and proprietary to common, to mention just a few extreme couplings."[11] Integrating digital elements will allow the built environment to become a connective tissue between the realities of bits and atoms—an interface that enables spatial cybernetics.

The built environment is becoming a physically habitable Internet, a Hertzian space—one that is inextricably intermeshed with digital devices. "Hertzian space is . . . a way of linking things, of sending information and content, etc. But [architecture] is an environment that can be inhabited, enjoyed, and explored."[12] In the newly interactive, digitally laced architecture, detail and dynamism and complexity (formerly the ambition of parametric scripting) are the experiential consequence of design, not the justification. Architecture takes on life through response—it becomes shocking or vibrant in time rather than in its fixed visual character.

Just as smartphones are a portal to larger systems, architecture can function as a mediator between daily, human-scale functions and vast, humanity-scale networks. "For millennia architects have been concerned with the skin-bounded body and its immediate sensory environment . . . Now they must contemplate electronically augmented, reconfigurable, virtual bodies

Digital Water Pavilion by Carlo Ratti Associati

Digital technology can vibrantly animate architecture in a way that is not at all contorted or blobby. Real-time sensors, actuators, and "spatial software" can create dynamic architectural experiences. The Digital Water Pavilion, shown here, began with a simple challenge: to use water—the theme of the 2008 World Expo in Zaragoza, Spain—as an architectural element. The design teams at the MIT Senseable City Lab and Carlo Ratti Associati sought to create dynamic, responsive, flowing architecture. The pavilion is a flexible and multifunctional space, with walls composed of falling droplets, each precisely controlled by digital nozzles to generate patterns, writing, or entrances. The result is a space that is interactive and reconfigurable. Using sensors, each wall responds when people approach, parting to become an entrance or an exit. Internal partitions can also shift depending on the number of people present.

that can sense and act at a distance but that also remain partially anchored in their immediate surroundings."[13] Pre-digital humans navigated their immediate physical surroundings, but today's cyborg (with prosthetic smartphone) inhabits space in profoundly different ways. Scales and contexts are blurred as we slip elastically between them. At any given moment, we may be standing in a room with three other people, but now the digital-spatial network can also reveal two close friends in a restaurant next door or a potential love interest only a block away. People and physical space are still a central anchor, but the upper and lower bounds of human reality have exploded outward, and architecture must encompass this breadth of spaces—in all of their active dynamics—while still relating to humans. Picon sets forth the question.

> How should the [designer] cope with an electronic
> and informational reality that seems to possess a
> dynamism and an expressive quality? . . . The advent
> of the digital represents an even greater challenge
> for design than what the early stages of mechaniza-
> tion had meant for modern architecture . . . For the
> first time perhaps, architecture has to confront itself
> with a deeply non-tectonic reality . . . Given these
> premises, how can the designer be in deep accord-
> ance with the invisible flows of information that
> constitute the bones and flesh of the digital world?[14]

The very process of creating architecture could become an iterative chain rather than a directly linear process. Today, design, documentation, construction, and inhabitation are distinct phases in the life of a building, each carried out by a different specialist using different tools. As each step of the architectural production chain transitions to digital systems, the whole process will be unified. Integration will happen incrementally, by streamlining information, enabling the different phases to inform one another, structuring a codependent feedback system and, ultimately, a full merger. Initial steps have been taken in this direction—for example, with project-specific smartphone apps that organize the fabrication, shipping, and installation of complex facades with tens of thousands of unique components.

Implicating inhabitants in all stages of the design, construction, and operation chain will graft the development and inhabitation of architecture into a single experience. People, digitally connected as an Internet of Bodies, and networked architecture will be symbiotic. "All evolution is co-evolution; individual species and their environments change and evolve on parallel courses, constantly exchanging information." What was formerly defined by a clear separation between mind, body, population, and environment is now entangled, "supplanted by a more complex and non-linear pattern of urban development in response to the spread of new information technologies."[15] Each choice we make has ramifications in digital space that, in

Vertical Plotter by Carlo Ratti Associati

Embedded technology can transform a building into a giant user interface—a system for collecting user input, changing the architecture, and displaying information or images. The world's largest plotter, debuted at the Milan World Expo in 2015, dynamically paints the facade of the Future Food District, a digital supermarket designed in partnership with the Coop Supermarket chain. The plotter is made of mechanical printer units that move along two axes and paint on the vertical surface with spray cans of different colors. The building's facade is transformed into a dynamic data visualization that reflects visitors themselves.

turn, shape our physical environment. The Internet of Bodies, grounded in our cyborg condition, may ultimately realize the concept of the built environment as a social and relational process.[16]

The most important implication of radically integrating digital systems into architecture will be to refocus technology and the built environment on humans. A living, cybernetic program in spaces of dynamic interaction will make architecture more like an extension of the body—and it is cyborg "tools" that enable the environment to respond. Augmented or "living" architecture is the large-scale hardware that digital-physical cyborgs create, plug into, and interact with. Active buildings are at once an environmental life support, a social catalyst, and a dynamic set of experiences. While congenital digital systems integrate seamlessly with human biology, the same prosthetic devices interface with the digitally augmented environment through real-time information flows. The Internet of spaces and the Internet of Bodies enable and co-create each other—each is the interface to the other. Ultimately, technology recedes into the background, and interaction is brought to the fore. Buildings can be simple—rather than voluptuous and shocking—but even more integrally vibrant and living.

PART III SENSEABLE CITY

*Forget the damned motor car
and build the cities for lovers and
friends.*
Lewis Mumford, 1979

SEVEN
MOBILITY

New cities of the machine age were animated at the speed of cutting-edge contemporary transportation technology: the automobile. Henry Ford's Model T, introduced in 1908, brought automobile ownership to the masses, and as adoption skyrocketed, cars had a profound impact on the fabric of cities. Convoluted networks of medieval or Victorian roads were eventually replaced by gleaming, organized superhighways designed for speeding car traffic. "The automobile is a new development with enormous consequences for the large city. The city is not ready for it . . . I tell you straight: a city made for speed is made for success."[1]

New transportation technology inspired radical visions of a new urban form, not only theoretical but also built. Brasília, a city designed by Oscar Niemeyer and Lúcio Costa and built from scratch, is a striking example of automobile urbanism. Conceived as Brazil's capital, the city was engineered to maximize speed and efficiency (and planned in the shape of an airplane, no less). Various urban elements—banking, hotels, embassies, and government buildings, for example—are kept separate, connected only by a network of highways. Most conspicuously, the city is without sidewalks or traffic lights; instead, intersections are enormous cloverleaf loops. Because there are (in theory) no pedestrians, there is no need for human-scale streets—people move through the city at the speed and scale of the automobile.

Brasília is a rare example of willful urban planning with a singular vision, but automobiles have transformed almost every city in the world—from brand-new, tabula rasa developments to historic city centers. Vehicle ownership increased rapidly in the early twentieth century, and cars quickly became an entrenched component of life and work. Urbanists saw the promise of enriched urban life and dove into a headlong rush to optimize cities for automobiles. In parallel, the increasingly popular car-based lifestyle exerted social, economic, and political forces. Cities were caught in a feedback loop: increased car ownership led to declines in public transit ridership, and simultaneously, policies and funds at the local and national level were diverted

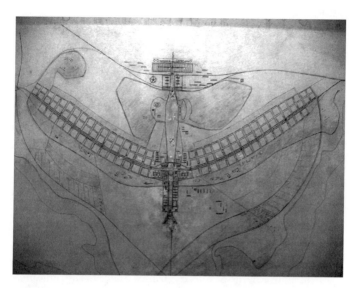

Plan of Brasília by Oscar Niemeyer and Lúcio Costa

Building a city entirely from scratch allows the planner to selectively use only the most advanced technology of the time. Recalling the long-standing race for urban efficiency, the masterplanned city of Brasília was designed in 1956 by two Brazilian architects and planners, Oscar Niemeyer and Lúcio Costa. The city is defined by state-of-the-art transportation technology—the automobile. (Seen from above, however, Brasília looks like an airplane.) Car culture dominates in a city composed almost entirely of highways. The original plan, shown here, contains no sidewalks or traffic lights, and different urban functions are separated into distant zones. The city is an important political and economic center, but it is almost without character or life, earning the city its nickname "ilha da fantasia," or fantasy island, in Portuguese.

away from public transit and toward highways.[2] Citizen behavior spoke clearly: more cars, more asphalt.

Schemes that targeted public transit exacerbated societal shifts toward personal mobility. What has come to be known as the "Great American Streetcar Conspiracy"—although the conspiracy remains unproven—choked public transit in cities across the United States during the 1940s and 1950s. A group of automobile companies, allegedly led by General Motors, implemented programs to purchase streetcar and electric train systems and subsequently dismantle them. The project was brought to the public spotlight by a whistleblower, Commander Edwin J. Quinby, in 1946, with accusations that there was a deliberate scheme to shift the United States toward automobile dependency. Although the companies were never legally prosecuted under antitrust regulations, the affair unambiguously contributed to the same vicious cycle: cities became increasingly hostile to pedestrians, and cars became increasingly necessary.[3]

The automobile became a symbol of the American dream, embodying success, individualism, and empowerment. A personal vehicle could satisfy any whim or fancy—unfettered by train schedules or bus routes, cars promised mastery of space and time. The allure of the automobile, particularly in mid-century America, was nothing short of pure freedom. The same attitude rapidly permeated—to varying degrees—most of the industrially developed and emerging world.

In almost perfect synchrony with the rise of automobile glamor Los Angeles sprang up out of the Southern California desert. With seemingly limitless space and wealth to match, the city spread itself from the ocean in the west to the Inland Empire in the east, resulting in a distinctive and disaggregated urban form. The pattern was so characteristic that the urbanist and architectural critic Reyner Banham made a pilgrimage from the United Kingdom to define and study it. He sought to understand not the signature buildings of the city but the urban fabric and its genesis—and to do that, he took to the roads. "Like earlier generations of English intellectuals who taught themselves Italian in order to read Dante in the original," said Banham in a colorful documentary, "I learned to drive in order to read Los Angeles in the original." What he found outside the windows of his car was a city built of four "ecologies": Surfurbia (the beach), Autopia (the freeways), the Plains of Id (the flatlands), and the Foothills. "The point about this giant city, which has grown almost simultaneously all over, is that all its parts are equal and equally accessible from all other parts at once."[4] Rather than a traditionally centric and radial urban form, Los Angeles spread in a cellular and vascular way, with each area interconnected through a tissue of road networks. In a society where everyone owns a car, every point is connected with every other, and the intervening space is irrelevant.

The extraordinary thrust of automobile optimism arced

through the first half of the twentieth century, but momentum inevitably waned. It became clear that unbridled car-focused urban development would have severe negative consequences. Another feedback loop had taken hold: the answer to more traffic is more roads, which, in turn, invite more traffic. Urban spaces spiraled out into sprawling suburbias that depended on the life support system of automobiles. The pattern was quantitatively described by the law of peak-hour expressway congestion, which mathematically demonstrates that "on urban commuter expressways, peak-hour traffic congestion rises to meet maximum capacity."[5] It follows, logically and empirically, that increasing road capacity can make traffic congestion worse, in addition to stifling public transportation:

> Almost before the first day's tolls on these expressways have been counted, the new roads themselves are overcrowded. So a clamor arises to create other similar arterials and to provide more parking garages in the center of our metropolises; and the generous provision of these facilities expands the cycle of congestion, without any promise of relief until a terminal point when all the business and industry that originally gave rise to the congestion move out of the city, to escape strangulation, leaving a waste of expressways and garages behind them.[6]

Automobiles nonetheless continued to define the twentieth-century urban development paradigm. With a goal of maximum traffic throughput, new highways cut through the built environment and fueled metropolitan sprawl. Consistent and passionate critics of suburbanizations, most vocally Lewis Mumford, held planners accountable, starting in the 1960s. Among Mumford's less subtle arguments is the iconic phrase "Forget the damned motor car and build cities for lovers and friends."

And yet, even today, urban spaces around the world continue to develop in the image of the American city. Urban planning is defined by car culture, and the resulting urban systems present few transportation alternatives. In some cases, the scale and severity of congestion are entirely unprecedented. In 2010, Beijing—a city notorious for its overcrowded ring roads—saw the longest recorded traffic jam in history: a blockage not caused by accidents, closures, or natural disaster but by the sheer number of cars on the road. At one point the stoppage reportedly stretched for sixty-two miles and incapacitated the highway for more than twelve days.

Traffic congestion has implications beyond throughput and delay. As cars idle, they continue to emit pollutants, releasing a maximum level of toxic emissions when they accelerate from a standstill. Crowded roads can cause acute spikes in smog, a pattern that is further exacerbated by certain geographic and atmospheric conditions: valleys that collect air, stifling summer heat, deep canyons between skyscrapers, lack of wind. A 2014 report from the World Health Organization states: "Few risks have greater impact on global health today than air pollution: the evidence signals the need for concerted action to clean up the air we all breathe." WHO estimates that every year poor air quality causes seven million premature deaths.[7]

The impact of automobiles resonates in a variety of less obvious ways as well—for example, parking. A high number of cars within city limits requires a proportional volume of parking infrastructure, and cities tend to naturally adjust the number of spots to satisfy peak demand. Parking availability escalates in much the same way as freeway capacity (demand rises to meet—and strain—supply). This situation has inspired compelling arguments against the unquestioned addition of parking infrastructure.

> Urban planners typically set minimum parking requirements to meet the peak demand for parking at each land use, without considering either the price motorists pay for parking or the cost of providing the required parking spaces. By reducing the market price of parking, minimum parking requirements provide subsidies that inflate parking demand, and this inflated demand is then used to set minimum parking requirements. When considered as an

impact fee, minimum parking requirements can increase development costs by more than 10 times the impact fees for all other public purposes combined. Eliminating minimum parking requirements would reduce the cost of urban development, improve urban design, reduce automobile dependency, and restrain urban sprawl.[8]

The public health threat of pollution and the infrastructural burden of parking are reaching broader awareness, but automobiles also have a less quantifiable impact on urban form and quality of life. Despite the best intentions of early planners, automobile-centric transportation systems, particularly at their present scale, are insensitive to the subtleties of urban space and, at worst, destroy the fabric of the city.

The answer to urban expansion and diffusion—and the host of social consequences that they bring—may be to optimize, rather than increase, transportation infrastructure. A first wave of developments started at the turn of the millennium, drawing on digital and physical systems. Top-down systemic engineering has been proven effective for achieving efficiencies in several cases around the world. Notably successful examples are electronic road pricing and flexible office hours. The first is similar to economic incentive programs that flatten peak energy loads by making power more expensive when it is in high demand:

when roads are crowded with commuters, the system responds by charging them more, effectively mitigating peak congestion. Various forms of Electronic Road Pricing have been implemented by cities around the world, including London, Singapore, Stockholm, and Milan, improving traffic in their downtown road networks. With similar intent, many corporations have introduced offset working hours to shift commute times earlier or later without impacting the duration of the workday.

There is also a bottom-up or decentralized response to readapting the network, one that instrumentalizes the existing infrastructure in an opportunistic way. Cars are idle approximately 95 percent of the time, making them an ideal resource for a sharing economy.[9] It has been estimated that every shared car can remove between ten and thirty privately owned cars from the road.[10] Zipcar, for example, puts a fleet of shared cars into the hands of a subscription-based community. Rather than each person owning a vehicle—using it perhaps twice a day and leaving it parked for the remaining twenty-three hours—a much smaller number of communal cars can satisfy the overall mobility demand.

Even more distributed peer-to-peer systems might emerge that allow the ride itself to be shared. Using a large dataset from taxi networks, a team of researchers at the Senseable City Lab has examined the potential impact of sharing car trips. They found that the mobility demand in several different global

HubCab by the MIT Senseable City Laboratory

The sharing economy is making inroads in transportation. More and more systems allow people to share cars: some are publicly funded, such as BlueIndy in Indianapolis, and others are based on private subscription services, such as Zipcar. Yet with pervasively networked platforms and real-time analytics, people may also be able to share individual rides. This simple hypothesis was the futurecraft scenario for a project called HubCab by the Senseable City Lab. A team of researchers created a mathematical model to determine the potential impact of ride sharing and applied it to a large dataset from New York City's taxi network. Shown here is a visualization of that dataset—all the taxi pickups and dropoffs in the New York area over the course of a year. Mathematical analysis demonstrates that 95 percent of trips can be shared and that the entire city's mobility demand could be satisfied by only 40 percent of the cabs in service today. The same numbers hold true for several different cities around the world and point to a near future in which innovative systems can cut travel time, costs, emissions, and traffic on our roads.

cities could be satisfied by only 40 percent of the cabs in service today.[11] Although the project developed a new mathematical model for "shareability networks," it was ultimately an act of design and futurecraft—imagining a future condition of wide-spread sharing, demonstrating the impact on vehicle use, and making the results available to the public—with the intent of opening possible avenues for development. Online platforms for networking and real-time data analytics could make this an immediate reality, connecting passengers and enabling trip sharing to radically transform urban mobility.

Such systems are dependent on local conditions, urban form, and social structures—for example, sharing systems would necessarily be very different in rural agrarian communities—but overarching trends are nonetheless evident. Even in sprawling suburban areas, real-time information can make public or shared transportation feasible, with algorithmically optimized mobility on demand. It would not be effective to plan a traditional bus route to a sparse community, but car or van sharing with real-time synchronization could be a viable option. Digital platforms stand to reactivate suburban areas and stymie the problematic feedback loop between cars, urban form, and social norms.

A host of emerging technologies are adding momentum to this trend. One advance that draws on sharing networks, data analytics, and hardware developments is self-driving cars. Autonomous mobility may be the final nail in the coffin of the individualist mobility paradigm, bringing the death of car culture but a rebirth of the (new) car.

The imminent generation of self-driving vehicles could be programmed according to a variety of different criteria, for example, comfort, fuel efficiency, or shareability. Self-driving could have tremendous impact at the urban scale, where telemetry and big data analytics might optimize vehicular flows through the city. Autonomous vehicles may prompt another wave of innovation in urban systems, from smart intersection management to procedures for dynamically rebalancing the vehicle network according to demand. For example, cars could autonomously migrate toward business centers at the end of the workday, preempting an increase in trip requests. As vehicles are increasingly shared, four out of five cars could be taken off the roads, and the remaining ones could be used in a more efficient way.[12]

The propagation of a new kind of urban infrastructure—silicon rather than asphalt—is eroding the symbolism of empowerment and emancipation that personal automobiles once carried. Previous attempts to reorient urban planning away from automobiles failed—not for lack of effort or sophistication but because the car was still firmly entrenched in daily life and culture.

A sea change is occurring today: the car no longer represents liberation. Individuals are empowered instead by a broad "transportation portfolio," a menu of options based on real-time information platforms that will ultimately enable a new regime of "ambient mobility." Personal transportation options are increasing in availability and sophistication, with an emphasis on shareability. Many cities around the world have city-bike and city-car systems, allowing visitors or residents to use a vehicle for a short period of time. An ambient mobility portfolio could also be tied to a constellation of external factors, from ecological footprint to personal health, including walking, running, or biking. Smart electric hybrid motors transform the bicycling experience and bring it online, while personal activity trackers show miles run, walked, or biked.

The freedom to choose between bicycling, sharing a car, walking, taking an on-demand taxi, using the subway or train, and hitching a ride with friends is far more appealing than owning and maintaining a car: it puts agency back into the hands of individuals. The trend is already apparent in drivers' license statistics—the percentage of young drivers obtaining licenses is diminishing sharply in the United States.[13] Generation Y has found a new way of using the asphalt laid by their parents.

A broader mix of mobility options can also increase efficiency, as the plurality of options allows the system to naturally balance itself. When information is delivered in real time—for example, "The bus is crowded and running slowly, so why not try a bike?"—individuals can make informed decisions, with a net positive effect. Not only will this activate unused capacity in the

REALTIME | MY CITY

Sun Dec 06 ◄   ► 03:47:50 | light rain

Copenhagen Wheel by the MIT Sense-able City Laboratory and Superpedestrian

For decades, cars have ruled the city, but a new generation of smart, networked transportation devices is taking hold. In addition to satisfying the urban mobility demand, these technologies can stream real-time information about the city and its environment. The Copenhagen Wheel transforms any ordinary bicycle into a smart electric hybrid. The red casing contains a motor, batteries, sensors, wireless connectivity, and an embedded control system. The wheel senses and learns how you pedal and integrates seamlessly with your motion, multiplying your pedal power between three and ten times. An array of onboard environmental sensors constantly collects data such as air quality measures, noise level, and traffic and routing information. Pictured is a visualization of data from an initial deployment of the wheel in Copenhagen.

transportation network, but additionally, it will empower the population to behave based on an understanding of the impact of each decision on overall urban function.

As ambient mobility platforms are widely adopted, public and private mobility paradigms will blur. What was formerly a clear (functional and social) delineation between shared and individual modes of transit will be erased. "Your" autonomous car can drive you to work and then drive someone else to school, rather than sitting idle in a parking lot all day. A single vehicle will go from one hour of use per day to twenty-four hours of use, as it is shared among a nuclear family, friends in a social network, a neighborhood, or an entire city.

Just as a small group of people might share an apartment, they might also share a set of mobility options. Social connectivity will become a key component of transportation strategies, aligning the number of vehicles with the number of travelers. This new structure will be compounded with improved intermodality, with the use of real-time information to streamline the transfer from one transportation system to another. Ambient mobility offers will integrate seamlessly, to the point of omni-modality. Commuters may bike to the station just in time to catch a train, and alight to find an autonomous car waiting for them at the station, ready to drive the last mile. Welcome to the age of the transportation portfolio.

*The fireside circle could no longer
serve as social glue. The old social
fabric—tied together by enforced
commonalities of location and
schedule—no longer coheres.
What shall replace it?*
William J. Mitchell, 2000

EIGHT
ENERGY

The earliest form of habitation technology was the grotto
—a natural feature that humans sought out for warmth,
protection, and sociability—and there they built the pri-
mordial hearth. Nomadic hunter-gatherer culture transitioned to a stable society, coalescing around the fire pit's
climate control system. For both sociability and efficiency,
shelter developed according to a centralized model. The
hearth was a focal point of social space—as the architect
Frank Lloyd Wright famously noted, it is the "psychologi-
cal center of the home."

Yet as time progressed, architecture's many dimensions
became decentralized, following an outward trajectory
of spatial liberation. What was once a circle of firelight

fractured into a proliferation of light fixtures in every room; the village well, formerly a site of gathering and gossip, flowed out through pipes to each home; even entertainment crossed the threshold of the theater and was beamed to cathode-ray tubes and screens in every living room. Elements of habitation are now individually and instantaneously delivered. Life is unmoored.

Climate control is no different—with the evolution of the hearth, heat was progressively liberated. Over time, humans exerted increasing control over temperature, until the "enforced commonalities of location and schedule" began to fray.[1] The Victorian era brought heat to homes, through pipes that circulated hot water. Each room could be temperature-controlled using iron radiators. The triumph of centralized domestic heat, half a century later, was the thermostat, a simple system that maintains a stable temperature at the desired setpoint by sensing ambient air and automatically turning central heating on or off.

Atomization, however, comes at the cost of efficiency— particularly in the case of climate control. The hearth is no longer a shared resource that attracts people but a distributed system in which each user demands the right to comfort at all times. With central heating and a binary on-off system, there has come to be a dramatic asymmetry between human occupancy and energy use. Entire homes are heated during the day when residents are at work or school, and even when they are home, empty corners of the house are indiscriminately kept just as warm as those in active use. To ensure constant comfort, we heat every space we might possibly inhabit.

Architecture could be conceptually reduced to a functional assemblage of environmental life-support technologies. Reyner Banham's 1965 essay "A Home Is Not a House" took a critical stance toward the modern domestic situation, suggesting a dissociation between environmental support and architecture. The essay begins with an incisive question: "When your house contains such a complex of piping, flues, ducts, wires, lights, inlets, outlets, ovens, sinks, refuse disposers, hi-fi reverberators, antennae, conduits, freezers, heaters—when it contains so many services that the hardware could stand up by itself without any assistance from the house, why have a house to hold it up?"[2] The project highlights our modern dependence on climate control technologies and the obsolescence of both social and natural environments. An image titled *Un-House Transportable Standard-of-Living Package* shows Banham and Dallegret sitting naked in a transparent "Environment Bubble" on either side of an air-conditioning unit.

The thermostat was invented to keep a constant ambient temperature at the user's discretion—and Banham called out the proliferation of such technologies and our concomitant dependence on them. Recent digitization, however, allows feedback systems that dynamically manage climate and allow technology to fall into the background without radically subverting the premise of architecture. Research shows that modulating energy usage based on occupancy could reduce consumption dramatically—in the case of the United States, by almost one-third.[3] One of the earliest devices, aptly named Nest, integrates smartphones with the home heating system. The digital thermostat learns from its users' daily habits, can be controlled remotely, and encourages various environmentally beneficial patterns, including some that are based on gamification and promote playful family dynamics. The evolved thermostat works together with occupants to optimize climate systems.

Nest dynamically adjusts temperature over time, but the next step could be a similar degree of control over space—that is, synchronizing heat with residents' physical location. In a future scenario of architecture that senses and responds, a dynamic system for *local warming* could enable fine-grained control over personal climates while simultaneously improving energy efficiency.[4] Using sophisticated motion tracking paired with dynamic heat emitters, an individual thermal "cloud" would follow each human throughout a building, ensuring constant comfort while minimizing overall heat requirements. "Man no longers seek heat . . . heat seeks man."

Sensor networks integrated with fine-grained response systems are beginning to save energy across a broad spectrum of habitation systems, not only climate control systems. In addition to directing energy where and when it is needed, the trajectory

Past and Future of the Thermostat: Examples from Nest and Honeywell Labs

The traditional thermostat, such as the iconic Honeywell dial depicted, maintains a constant temperature at the users' discretion, but at the expense of efficiency. Entire homes or offices are kept comfortable, even if they are entirely or partially empty. Yet digital integration is causing rapid transformations in climate control technology. Nest represents the next generation of home climate system, one that aligns heat with daily and seasonal rhythms. It self-adjusts and builds personalized schedules by integrating directly with smartphones—even warming up a home before its residents arrive if they decide to come home earlier than usual. A digital control system dynamically modulates temperature based on the patterns it learns from occupants, improving overall energy efficiency.

toward sustainability is progressing in another way: mitigating peak loads.

City dwellers tend to demand energy at the same time (for example, at 7 p.m.), so to ensure that lights illuminate when any person (or every person) flips a switch, power plants must constantly produce enough energy to satisfy the maximum possible

Local Warming by the MIT Senseable
City Laboratory

A staggering amount of energy is
wasted on heating offices, homes, and
partially occupied buildings. Energy
is used to change the temperature of
empty air, rather than the temperature
of people themselves. Local Warming
addresses this asymmetry by synchronizing climate control with humans.
Responsive infrared heating elements,
guided by sophisticated motion-tracking
sensors, are mounted around a room.
These emitters can transmit collimated
heat to create a precise personal (and
personalized) climate for each occupant.
Individual thermal clouds follow people
through space, ensuring constant
comfort while dramatically reducing overall energy use. Pictured
is an early prototype of Local
Warming.

demand. Pattern analysis and predictive models can help align supply to demand, but even so, it is difficult for plants to tailor production effectively.

In 1981, Buckminster Fuller put forward the radical concept of a Global Energy Grid, a worldwide system of electroducts for international energy transfer. This would enable different countries to balance each other's supply and demand: when Europe is demanding energy during the day, China is asleep, and vice versa.[5] Furthermore, at any given moment, one side of the globe is facing the sun and could potentially be harvesting solar energy. In theory, the Global Energy Grid would send power from the sunny regions with surplus energy to dark regions with a deficit. The crux of Fuller's plan was reducing variations in the global system: mitigating the peaks and valleys. Fuller summarized the concept in characteristically sweeping terms:

> I have summarized my discovery of the option of humanity to become omni-economically and sustainably successful on our planet while phasing out forever all use of fossil fuels and atomic energy generation other than the Sun. I have presented my plan for using our increasing technical ability to construct high-voltage, superconductive transmission lines and implement an around-the-world electrical energy grid integrating the daytime and

nighttime hemispheres, thus swiftly increasing the operating capacity of the world's electrical energy system and, concomitantly, living standard in an unprecedented feat of international cooperation.

Global demand would dictate energy transfer, sparking what Fuller believed would be a shift of the economic standard from gold to kilowatt hours. "Such intercontinental network integration would overnight double the already-installed and in-use electric power generating capacity of our Planet," Fuller concluded.[6] The idea carried remarkable implications for sustainability, economy, and society.

Although the idea of a global superconductor network is alluring, it remains technologically and financially challenging. However, optimizing existing systems from the individual to the urban scale might achieve similar ends. Today, built environments are beginning to dynamically respond to humans in real time using sensing and actuating feedback loops. These responsive digital systems may control energy generation, demand, and distribution. The behavior of these dynamic systems "changes over time, often in response to external stimulation or forcing." The term "feedback" refers to "a situation in which two (or more) dynamical systems are connected together such that each system influences the other and their dynamics are thus strongly coupled."[7] As these systems blanket our cities, every dimension of habitation can be transformed, from the simplest example, occupancy-sensing lights in a single room, to complex systems for sensing, modulating, and optimizing energy patterns across an entire city.

According to the United States Department of Energy, "We are stretching the patchwork nature [of the existing electric grid] to its capacity. To move forward, we need a new kind of electric grid, one that is built from the bottom up to handle the groundswell of digital and computerized equipment and technology dependent on it—and one that can automate and manage the increasing complexity and needs of electricity in the 21st Century."[8] This is the promise of the *smart grid*.

In a very basic sense, the smart grid is simply an introduction of dynamic control systems for energy production, distribution, and consumption. The concept is rooted in an infrastructural framework of distributed (preferably renewable) energy production. With an integrated digital control system at the neighborhood or regional level, each house could generate energy and share surplus with others nearby or store it in local batteries. Today's archaic energy-switching technology will transition to a digitally controlled system, allowing faster response to real-time conditions.

Smart devices for end users can dynamically configure their consumption patterns based on information from the grid—a refrigerator, for example, can cool when energy is inexpensive and cycle off during peak demand. But the system is not exclusively automated top-down control. It also enables bottom-up incentivized response. Networked smart meters stream real-time information, monitor local and regional demand, and offer incentives directly to users. Surge pricing—that is, pricing that changes in response to demand—provides a financial incentive for users to conserve resources. Individuals are free to make decisions and can do so with the knowledge of overall energy demand. In more domestic terms: your refrigerator might automatically adjust its on-off cycles for efficiency, but running the dishwasher or charging a computer are still individual choices. Smart meters can inform real-time dynamic pricing so that people are free to use power however they want, but when demand is high, the cost will rise. By whatever means, the goal of the smart grid is to mitigate and level out peaks in demand and to reduce the amount of power generation required.

A truly functional smart grid is still quite far in the future, yet it is already possible to implement more efficient energy systems. The likely interim step will be a hybrid system situated between local and regional production, one that incorporates a wide array of urban infrastructures, from architectural batteries to systems for using cars as accumulators. Efficiency will be achieved by centralized distribution systems, optimized by responsive feedback loops, and incrementally supplemented with organically growing local production and consumption networks. If the two can balance dynamically, the marriage of complementary systems would allow for large centralized energy harvesting to fill in the gaps of local production grids.

This points to a future in which the overall energy infrastructure is dynamically managed, as each kilowatt-hour package is tagged and carries a variable price based on real-time supply and demand. Energy will be directed and delivered with intention, precisely where it is needed. In a near future, every device—and every vehicle and every building—will transfer energy in and out, constantly communicating with the broader network to balance overall system flows. Energy supply will respond to demand: the network itself will mitigate peaks and dips as it interacts with human dynamics.

The factory of the future will focus on mass customisation—and may look more like those weavers' cottages than Ford's assembly line.

Paul Markillie, 2012

NINE
KNOWLEDGE

The first industrial revolution profoundly reconfigured society. Beginning with Britain's iron and textile industries, innovations in factory procedures and powered machine technology sparked mass production. The shift from hand to machine fabrication brought about a profusion of factories, which, in turn, required an expanded laborer class. A host of unskilled workers executed highly specific tasks in the long chain of fabrication while a small demographic of intelligentsia orchestrated the process, and an even smaller elite reaped the benefits of the system.

There was a sharp demarcation between productive (repetitive) and intellectual (creative) work, "a transmuted form of the barbarian distinction between exploit and drudgery," so to speak.[1] People were reduced to functional components of a larger system that was itself mechanical: countless hands hovering over conveyor belts executing repetitive tasks in identical factories for endless hours.

The significance of individuals and their talents diminished as humanity acquired value only in numbers. Lewis Mumford offered an incisive summary: "We have created an industrial order geared to automatism, where feeble-mindedness, native or acquired, is necessary for docile productivity in the factory; and where a pervasive neurosis is the final gift of the meaningless life that issues forth at the other end . . . By his very success in inventing labor-saving devices, modern man has manufactured an abyss of boredom that only the privileged classes in earlier civilizations have ever fathomed."[2]

Not only did the industrial revolution reshape social structure, it also radically respatialized cities. Prior to the late eighteenth century, craft production took place in the residential workshops of fairly isolated villages. But as society shifted its gears for maximum output, the formerly agrarian population flooded into cities, looking for work. A new urban typology emerged: cities expanded into distinct zones for production (factories) and habitation (mass housing). The influx of workers exceeded the rate of expansion, and in crowded centers such as London, Manchester, and Birmingham, conditions for the working class were dismal.

The challenge of spatial optimization for production and housing inspired some early experimentation with entirely new factory towns in the years preceding the industrial era. These were engineered as meticulously as the production lines they hosted, which were optimized for throughput. One of the archetypes of the master-planned city from the proto-industrial era was the French Royal Saltworks at Arc-et-Senans, by Claude Nicolas Ledoux. In both organization and decoration, the saltworks complex expressed the supremacy of human rationality: drawing on ideas about the natural structure of the universe and geometric mathematics, the Royal Saltworks were a crystallization of contemporary French society on the brink of industrialization. Architecture was at once physics and metaphysics, orchestrated for production. The plan, for example, is a hemisphere, representing geometric purity as well as providing optimal visual access to the overseer and maximizing the number of living units with direct access to work areas. Ledoux understood the facility as two interdependent systems and two geometries: the administrative directorship, including the overseer and the tax agents, was organized linearly on the diameter of the hemisphere, and the workers' housing was arrayed radially on the perimeter.

The project was followed by a host of similar production-cities. Urban form was a spatial expression of the output-oriented social structure. The epoch's momentum continued, and "la ville

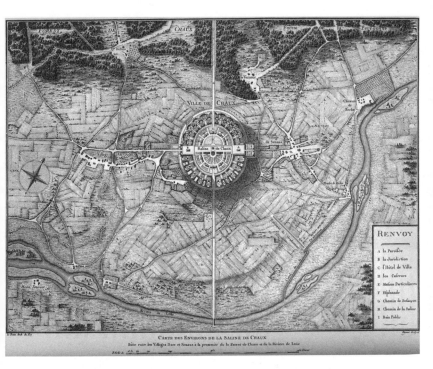

CARTE DES ENVIRONS DE LA SALINE DE CHAUX
Batie entre les Villages Darc et Senans à la proximité de la Forest de Chaux et de la Rivière de Loue

Royal Saltworks at Arc-et-Senans by Claude Nicolas Ledoux, 1775–79

There is a long tradition of using spatial planning, at the architecture, campus, or urban scale, to promote mechanistic productivity. The French Royal Saltworks at Arc-et-Senans were master-planned in the 1770s as an expression of both social and functional ideals. The plan, shown here, reflects Enlightenment-era philosophy—the geometric position of humans in the cosmos and the rela-tionship between overseers and employees—and the emerging economic reality of the factory city. Different elements were arranged to optimize the workers' daily tasks and reinforce the factory's human hierarchy. The campus was an expression of contemporary French society: the triumph of rationality and an economy poised for industriali-zation.

fonctionelle" reached a pinnacle of spatial orchestration at the merger of architecture, labor, and society. State-of-the-art transportation systems were imagined to link single-use urban zones for labor, habitation, and leisure. These functional utopias were designed to raise the standard of living for each employee while maintaining maximum productivity.

This mechanistic approach to urban form was a continuation, even an expression, of the industrial-era labor mentality. As manufacturing technology became increasingly sophisticated, the pace and precision of fabrication processes rose ever higher. The entrepreneur and inventor Henry Ford—a key figure of the so-called second industrial revolution—orchestrated meticulous production lines for low-cost, high-output fabrication. The epochal Ford automobile facilities churned out cars at an unprecedented rate: one every three minutes. The new methods increased production eightfold, reducing the number of labor-hours per car from about 12.5 to 1.5. Le Corbusier visited the Detroit factory and was so impressed by its streamlined operations—by what was considered to be the future of fabrication, industry, and architecture—that he reportedly exclaimed, "I am immersed in a type of astonishment!"[3]

Despite advances in machine technology and procedural configurations, however, humans were still chained to the factory line. Throughput skyrocketed, ushering in the era of mass production, yet working conditions were still defined by long hours, physical danger, low wages, and repetitive tasks.

At that time, the vision of ideal production was a future in which humans—prone to errors, delays, and strikes—were incrementally engineered out of the factory line and replaced by automation. The idea that "Mechanization Takes Command" (a phrase coined as the title of an iconic book of the time) proposed a new kind of interaction between man and machine. That is, "the problem of the assembly line is solved when the worker no longer has to substitute for any movement of the machine, but simply assists production as a watcher and tester."[4] With a series of technical innovations, manual tasks could be left to machines, and human labor could shift away from monotonous repetition —or even cease altogether.

Complete automation could, theoretically, relieve humanity of all labor obligations, triggering a societal shift from production to play. The term *Homo Ludens,* coined by the cultural historian Johan Huizinga in 1939, refers to this hypothetical phase in so-cial evolution. "Modern fashion inclines to designate our species as Homo Faber: Man the Maker . . . It seems to me that next to Homo Faber, and perhaps on the same level as Homo Sapiens, Homo Ludens, Man the Player, deserves a place in our nomen-clature."[5] Play (as distinct from work) can be understood as the primary impetus and expression of human culture, the force that creates and animates society. If fabrication and production are outsourced to machines—and adequate equity measures govern access and control of technology—then play could be the last and greatest human activity.

Since the beginning, humans have had to occupy themselves with survival, but some theorists have imagined that the demands of time and effort could diminish and even vanish. This shift would have an even more transformative—and diametrically opposed—impact on society than the industrial revolution. The Dutch artist Constant Nieuwenhuys based New Babylon—a decades-long project for social, aesthetic, and urbanistic exploration—on this premise. "The opposite of utilitarian society is ludic society, where the human being, freed by automation from productive work, is at least in a position to develop his creativity . . . it is clear that a ludic society can only be a classless society. Social justice is no guarantee of freedom, or creativity, which is the realization of freedom. Freedom depends not only on the social structure, but also on productivity; and the increase in productivity depends on technology. 'Ludic society' is in this sense a new concept."[6]

Though grounded in very real technological developments, Constant's New Babylon and other such projects offered a speculative future that failed to materialize. Yet the widespread adoption of digital fabrication technology is restructuring production in different ways—spatially, procedurally, and socially.

These developments have been branded with an iconic label: the third industrial revolution.[7]

Three main transformations are taking place. First is the possibility of creating material forms through digitally controlled additive processes—that is, by laying precise deposits of material to build up a shape—using 3D printers. Not only does this allow for much more complex geometries than have ever been possible, but it also shatters the established laws of mass production and economies of scale. Industrial-era factories churned out large quantities of identical objects, reducing cost through repetition. According to that model, a bespoke item—say, a customized Rolls-Royce—was extremely expensive. For 3D printing and digital fabrication, on the other hand, there is effectively no difference between creating identical versus unique objects. Items can be manufactured for about the same cost, whether by the thousands, by the hundreds, or for a single unit. This is a complete reversal of the Fordist factory lines that churned out identical products according to the mantra "You can have any color car, as long as it is black." Digital fabrication will usher in an era defined by individual control. "The factory of the future will focus on mass customisation—and may look more like those weavers' cottages than Ford's assembly line."

The second transformation is the possibility of fluent transition from digital outputs to physical objects. Thanks to subtractive CNC machines (computer-controlled machines for drilling, cutting, carving, and more), and additive 3D printers, digital code can become physical material or action with a click. In much the same way as personal printers allowed people to create documents in their homes, the production of *things* is quickly becoming customizable and immediate. As the boundary between software and hardware is blurred, custom fabrication will be carried out on demand. The act of making objects will become more like compiling and executing code than like laborious, specialized, and time-intensive woodcraft in a carpentry studio.

The third transformation—a result of the fluency between digital and physical—will be social. Using intuitive software, anyone can create and upload a design online to be shared with friends, communities, or the public at large. Just as in open software, a project itself could spark new modes of collaboration between a variety of actors. The architect David Benjamin writes, "It's much easier to use [digital] tools and the equipment is cheaper, so the projects are getting more interesting. But most importantly, the community around these projects has grown: people do a project, publish their process and results, and then other people ask questions about how it was done and discuss the project. Once there's that community of people sharing projects with an open source ethos, that's kind of unstoppable. It's not really the technical stuff; it's the social stuff."[8] A marketplace of downloadable and printable objects could displace or

Fab Lab by Applied Nomadology

Education, traditionally, is a one-way flow of information from teacher to student. Yet tools for building and making invite anyone to create knowledge through personal experience and to diffuse that knowledge through networks of peers. Fab Labs put tools for building and making into people's hands, inspiring creativity and community. A global network of these spaces, like this one in Amsterdam, provides citizens with unprecedented access to tools, inviting them to create anything and everything they can dream of. Fab Labs harness the power of networks—open sourcing, digital design, and social media—and enable the compelling experience of fabrication, as people see their ideas become reality and create something tangible and useful. The new pedagogical model is based on the idea that people learn much more effectively if they engage with something personally meaningful rather than passively absorb ideas.

redefine professional designers through an alternate economy driven by either financial or social transaction. The fabrication process itself could happen domestically, in individual homes—if 3D printers become as ubiquitous as inkjet printers—or in neighborhood-level fabrication facilities.

Building a worldwide network of local communities around neighborhood fabrication facilities is the vision of Fab Lab, a program that began at MIT. Since the doors of the first Fab Lab were opened in 2001, new shops have cropped up around the world, from campuses to inner cities to rural villages, offering tools for digital and physical fabrication. The projects coming out of them have a local inflection, as communities come together to solve problems and generate new ideas—at a Fab Lab in Norway, for example, shepherds put together radio-frequency ID tags for tracking wandering sheep. The founder of Fab Lab, Neil Gershenfeld, explained the idea and genesis of the project in a TED presentation. "Instead of talking about it, I'd give people the tools. This wasn't meant to be provocative or important, but we put together these 'Fab Labs.' And they exploded around the globe . . . The real opportunity is to harness the inventive power of the world, to locally design and produce solutions to local problems."[9] This is a new form empowerment—Fab Labs allow people to modify or "hack" the world around them, rather than passively absorbing information and products. As people design and construct technology themselves, it becomes localized, instrumental, and practical.

Fab Labs are places not only for production but also for learning. Crucially, each lab is the nucleus of a fabrication-focused community. Many labs host weekly classes, workshops, and social events. "The message coming from the Fab Labs is that the other 5 billion people on the planet aren't just technical sinks, they're sources," and they are propelled by a new possibility of merging education, experimentation, and making.[10]

The first two industrial revolutions reshaped cities, and today's decentralized fabrication might have no less profound implications for urban form. Production could be realigned with daily life as manufacturing exits the factory. Society could return to a preindustrial model, one that is local and user-centric—and futurecraft can be applied to guide the changes. New domestic typologies for the twenty-first century might recall medieval cottages in Great Britain, Peranakan shop houses in Singapore, or *machiya* in Kyoto's artisan districts, combining dwellings with fabrication.[11] If not in individual homes, a dispersed urban platform for community fabrication activity may be spread throughout the city, establishing an open infrastructure that turns community members into makers and becomes the center for sharing knowledge, creating, and socializing.

This vision is the ultimate capitulation of industrial-era

zoning. The city fabric would be reconstituted as the workplace and the home collapse into a hybrid unit and as a more social, community-based model blurs formerly distinct urban districts. The city may come to life in new ways. "One potential outcome of all this, where zoning and other policies allow it, is a clustering of the new-style live/work dwelling in twenty-four-hour neighborhoods that effectively combine local attractions with global connections. These—not isolated, independent electronic cottages—will be the really interesting units in the twenty-first-century urban fabric."[12] Not only will design and production respond to local conditions in a sustainable and targeted way but the city will become more livable. Spaces of human habitation will become functionally intermixed to the point of being broadly homogeneous yet vibrantly active. When the factory is everywhere, cities will be productive on a fine-grained (human) scale.

PART IV LOOKING FORWARD

The city, messy, anarchic, . . .
has been a place where those
without power get to execute a
project. They get to make history.
Saskia Sassen, 2013

TEN
HACK
THE CITY

Over the course of the 1990s a new form of power
emerged—one that reverberated on the global scale . . .
and could be wielded by a scruffy teen in a basement. Film
depictions like *The Matrix* and *Sneakers* dramatized cyber
security and the power of connected computers as they
became a very real force in the world. A new figure was
cast: the hacker.

The notion of a hacker initially emerged in the domain
of computer science, defined as "an expert at program-
ming and solving problems with a computer; a person
who illegally gains access to and sometimes tampers with
information in a computer system."[1] Essentially a person
who claims power in digital space, the hacker captured

public imagination as the world migrated online during the 1990s. Huge global networks were being built, and their (malicious) destabilization became an immediate possibility. Hackers' activity was wrapped in a shroud of mystique, surrounded by call signs, offshore teams, and infected USB drives.

Hacking into information systems, however, is nothing new—it goes hand in hand with the emergence of telecommunications. One of the first attacks struck Guglielmo Marconi's demonstration of radio transmission in 1903, when he communicated from Cornwall to London, three hundred miles away. Nevil Maskelyne—a music-hall magician and would-be wireless tycoon—had been frustrated by the Italian inventor's patents. In what was either an act of aggression or a demonstration of prowess, Maskelyne managed to take control of the system and broadcast obscene messages to the Royal Institution's scandalized audience.

With the spread of digitization during the twenty-first century, more and more systems can now be hacked. In the era of ubiquitous computing almost everything can be accessed, appropriated, and subverted—from computers to cars to washing machines. As the world goes online, it is becoming a hacker's playing field.[2]

The scope of "hack value" has simultaneously expanded—it is certainly more than creating malicious computer viruses. "It is hard to write a simple definition of something as varied as hacking, but I think what these activities have in common is playfulness, cleverness, and exploration. Thus, hacking means exploring the limits of what is possible, in a spirit of playful cleverness. Activities that display playful cleverness have hack value."[3] Generally speaking, hackers are motivated by a challenge, and there are more and more challenges in hybridized digital and physical space.

The concept of "cleverness" is somewhat motive-agnostic, with connotations ranging widely, from innovative to sly to malicious. Although the tools and procedures of computer hacking are largely universal, the end results can be benign or malignant, creative or destructive. Two categories of hacker are generally identified as "white hat" and "black hat." The former is an ethical hacker who seeks to penetrate a computer system and identify its flaws in order to make it more secure. The latter is a hacker who "violates computer security for little reason beyond maliciousness or for personal gain."[4] Black hat hackers have been codified as a new class of felon in US legislation, categorized as "cyber-criminals."

The 1990s saw a flurry of hacking activity on both sides of the white hat–black hat boundary. Cases spanned a wide spectrum, from cracking security systems to artificially winning a Porsche in a radio-show giveaway. Adrian Lamo became one of the most prominent figures in hacker culture, notoriously breaking into corporate systems such as Excite@Home, MCI WorldCom,

Yahoo, Microsoft, and Google. Most of the trespasses were magnanimous—Lamo subsequently contacted the companies, pointed out security flaws, and suggested fixes. His pseudo-legal activity finally caught up with him with a grand hack of the New York Times Company, where he uncovered the personal details of thousands of contributors to the paper, including celebrities and former presidents. To avoid jail time, he negotiated a plea bargain that included six months of house arrest.

Kevin Poulsen was a hacker from an early age. As a child, he learned how to whistle into a pay phone to connect a call for free, and over time his challenges incrementally grew in both risk and reward. Poulsen famously monopolized phone lines to ensure that he was the 102nd caller to a Los Angeles radio show and won a Porsche car. His career on the wrong side of the law climaxed in an extended chase with the FBI followed by fifty-one months of imprisonment. He was the first American ever to be sentenced with a ban on computer access after his release from prison (effective for three years).[5] After serving time, Poulsen doffed his black hat and became a senior editor at *Wired* magazine, focusing on cyber security.

Hacking is difficult to define, particularly because the boundary between malicious and magnanimous is blurry. The consistent and unifying feature is a common culture—one that very much predates the computer. Much of hacker culture originated on the MIT campus with the Tech Model Railroad Club. With a stated passion for discovering the way things operate, mechanical and otherwise, the founding members shaped what would become hacking culture, codifying their ideas and vocabularies in the 1959 *Dictionary of the TMRC Language* (later a bible in geek lore). In this document they coined such phrases as "information wants to be free" and gave us the very term "hacker"—originally referring to someone who pulls off an elaborate college prank.

The club's first home was on the third floor of MIT Building 20, the "Plywood Palace," which was in many ways an incubator of early hacker aesthetic and ethos. The group was concerned with far more than model trains: they first subverted and reconfigured the room itself, as the history of the club makes clear: "Having space in a temporary building had the advantage that no one really cared what you did there, including altering the wiring or modifying the walls."[6] Building 20 was fertile ground for nearly limitless experimentation with both content and context, as it was for many other groups and researchers. The emerging aesthetic was ramshackle and ad hoc, but the building was a petri dish for (often anti-authoritarian) ingenuity.

There is a long history of appropriating and controlling physical space as a platform for ideological engagement, stretching far beyond the MIT campus. The city is a venue, and its plazas and squares can be loudspeakers. Because civic space is so often charged with a specific social, political, or religious identity, hacks in that space can be correspondingly incisive. Almost every revolution has begun and ended in a public square: from Julius Caesar entering Rome to the 2014 protests in Ukraine. In the epochal moments of history, public actions subvert the intended use of the space or the ruling party that lays claim to it. The Place de la Concorde, the largest public square in Paris, was originally intended as a signifier of the monarch's power and was therefore a fitting site for the execution of Louis XVI. On January 21, 1793, at the culmination of the French Revolution, the king was beheaded, and the square was renamed Place de la Révolution. More than two hundred years later, Tahrir Square in Cairo saw mass participatory action by citizens that resulted in the downfall of Egyptian presidents Mubarak and Morsi in 2010 and 2013, respectively.

Civic action in physical space is not exclusively negative or revolutionary, nor must it be targeted at a specific leader. The city is simply a platform and an amplifier to be appropriated (or hacked) for any purpose. From the breadth and variety of uses— or hacks—throughout history, places acquire unique mythologies as aggregate residue. The National Mall in Washington, DC, one of the most visible spaces in the United States, has broadcast voices nationally and internationally: the "I Have a Dream" speech by Dr. Martin Luther King Jr. in 1963, the Vietnam War Moratorium Rally in 1969, a mass with Pope John Paul II in 1979, concerts by a wide variety of musicians, from the Beach Boys to Britney Spears, and the presidential inauguration every four years. While these events are not hacks in and of them-

People Congregated in Tahrir Square on February 9, 2011, by Jonathan Rashad

Ideas spread quickly online—social media, forums, and communication tools are all powerful media for community discussion, debate, and action. But physical urban space is, and will remain, the final stage. Revolutionary ideas and sentiments can spread like wildfire through online communities and spill into physical space, effectively challenging governments with unprecedented strength. A series of political demonstrations across the Arab world during 2011 have become known, collectively, as the Arab Spring. Strikes, marches, and protests were organized—and communicated to a rapt global audience—via social media platforms like Twitter and Flickr (the source of this photo). Pictured here is Tahrir Square, Cairo, during an event that has become emblematic of the movement. Through this demonstration, and series of similar ones, protestors contributed to the change of regimes in Egypt.

selves, they have a cultural weight that aggregates and lends gravity to future actions in the same space. Hacking the city in such a culturally charged site draws on the legacy of appropriations of civic space.

These events and actions prove the strength of urban hacking as a language of social statement. An iconic example by the artist and activist Oliviero Toscani, working for the Benetton group, radically challenged the Catholic culture of France and raised global awareness on World AIDS Day in 1993: the artist placed an enormous pink condom (seventy-two by eleven and one-half feet) on the obelisk in the Place de la Concorde. Given the immediate scrutiny of media outlets around the world, Toscani's act brought unprecedented media visibility to the disease, to social stigmatization, and to sexual health. In the same

Pink Condom by Oliviero Toscani for United Colors of Benetton, December 1, 1993
Public spaces can serve as tremendous platforms for social commentary. Cities are familiar—streets and squares and monuments belong to each and every resident—which lends power and gravity to urban subversions. As an undercover project for the 1993 World AIDS Day, the artist Oliviero Toscani installed a giant pink condom on the Luxor Obelisk at the Place de la Concorde, creating a shocking symbol of AIDS awareness in the center of Paris, at one of the city's most visible locations. Authorities intervened almost immediately, and the condom was removed the same day. Yet it captured the public imagination—since then, the hack has been repeated dozens of times, bringing a remarkable degree of visibility to the AIDS cause.

Paris, December 1st, 1993
World AIDS Day

city in 2014, ten thousand farmers protested European Union agricultural policy reforms by marching sheep through the streets, even taking them into the Louvre. The argument—that farmers do not make high-enough wages to take time to enjoy national treasures like the Louvre—was made blatantly clear as sheep flooded the museum's hallways. But even small hacks can make a sharp social critique, such as slightly altered signs in the London Underground that point out—and seek to improve—commuters' social habits by printing phrases like "Warning: do not make eye contact with fellow passengers."

The hack value of these instances is the connection between the particular space—its intended purpose and its history—and the revelatory transformation of the hack itself. Urban interventions would mean nothing without the physical settings they engage and without public scrutiny. If hacking is about understanding a system, appropriating it, and using it for alternate purposes, then the core of a truly successful hack in urban space involves, first, what the site means; second, how the hack appropriates the site; and third, how the hack transforms the site to communicate a message to a broad public.

Today, at the convergence of bits and atoms in urban space, there is a new dimension to hacking. Cities are increasingly suffused by a networked digital layer, and three important elements of hacker culture—computer geeks, protests, and social critiques—are powerfully colliding. What are the means, the significance, and the potentials of hacking in tomorrow's smart cities? The digital and physical juncture will play host to—even amplify—the good, the bad, and the ugly of hacking.

The bad is already familiar. In the wake of notorious hackers like Lamo and Poulsen, the potential for destabilizing multinational systems through cyber terrorism has become a palpable threat—one that becomes yet more menacing as physical infrastructures are digitized. Metropolitan-scale systems such as subways and gas lines are immediately vulnerable to antagonistic infrastructural hacking.

Malicious weaponization of infrastructure systems has been a common subject of recent fiction and film. The threat resounds dramatically as a plot line precisely because it is uncomfortably close to reality. In 2010, Stuxnet, a powerful (and very real) malware, wiped out over one-fifth of Iran's nuclear centrifuge facilities. The virus targeted industrial-scale programmable logic controllers—the same kind of real-time control system drivers that are behind factory machinery, amusement parks, and digital light fixtures. Stuxnet came to life almost biologically: it self-propagated from machine to machine, it covered its tracks by sending all-clear messages from infected systems, and it lay dormant if its execution criteria were not met on a given machine. When active, the virus overrode the control system that governed a centrifuge's massive rotating components, causing the facility to violently tear itself apart. The elegance and sophistication of the malware is often cited as incontrovertible evidence that major governments were on the playing field. Stuxnet took cyber-security specialists off guard. From their perspective: "We're all engineers here; we look at code. This was the first real threat we've seen where it had real-world political ramifications. That was something we had to come to terms with."[7]

Stuxnet proved the power of hacking at the juncture of digital and physical systems, and as the city becomes increasingly hybrid, the danger increases. The destructive potential of appropriating almost any smart system is proportionally threatening. Every dimension of the city could potentially be hacked: imagine careening autonomous vehicles, aggressively responsive architecture, or menacingly asynchronous energy distribution. A black hat hacker wields enormous power in cyber-physical space. In 2012, the US defense secretary, Leon Panetta, warned that the United States was vulnerable to a "cyber Pearl Harbor" that could, as a reporter paraphrased, "derail trains, poison water supplies, and cripple power grids."[8]

Hacking can serve an important positive function, however, both for social reasons and for security. The team at Kaspersky Lab that finally tracked down the Stuxnet virus (and its equally malicious progeny) in 2010 was a crack squad of former hackers and computer scientists experienced with tools and methods on both sides of the legal boundary. Such familiarity provides an advantage in analyzing and diagnosing existing systems and in designing tighter security.

By exploiting loopholes and executing hacks, white hat hackers can discover vulnerabilities and point toward new standards, security protocols, or encryptions to increase safety. This is the potential of hacking as innovation. Just as the shabby—that is, hackable—MIT Building 20 was a hotbed of discovery, so, too could tomorrow's smart city spark creativity by incorporating a measure of hackability. Many people have advocated for more hackable cities in a variety of ways. Given a platform—the urban analogue of iOS or Android, for example—anyone and everyone could create city apps. One admittedly low-tech example is Parking Day, an opportunity for citizens across the country to appropriate parking spaces, design interventions, and suggest alternate futures for the enormous—and largely wasted—parking infrastructure. Since its inception in San Francisco in 2005, Parking Day has expanded across the globe through the use of digital platforms.

But this kind of initiative brings up an important consideration: the delicate balance between inspiring and pedantic. That is, if you make something hackable, have you neutered its hack value? If it is no longer a challenge, will it be hacked?

On the urban scale, an immediate answer is to open both platforms and data. Urban information would be fertile soil for innovation but nonetheless hands-off from a top-down perspective. Promoting openness could "activate additional elements of both knowledge practices and technological practices, generate more engagement by city residents, more cross-neighborhood comparisons, scale up to city level but from the ground up, lead to exchanges and collaborations, and on to a fully mobilized neighborhood and city culture." The city may become a complex mesh of transparent systems, where a rich trove of information promotes appropriation and experimentation. Access may come to be understood as a basic human right, what the philosopher Henri Lefebvre called "le droit à la ville," the right to the city. It is the demand for a "transformed and renewed access to urban life."[9]

Open data and platforms are nothing short of a key to sharing ideas, knowledge, and best practices with willing collaborators. "This would be but one step in a potentially much longer trajectory, one that might entail a re-making of the urban." Open data and platforms are as valuable to a smart, living city as accessible and open public space is to the traditional city. The processes of urbanization are "fundamentally political questions" that can only be realized between the top-down and bottom-up.[10] Open place–based application program interfaces (APIs) or data repositories may become new forums to spread ideas, to critique standard practices, or to express sociopolitical views.

Tools for opening cities will fulfill and make tangible a long-standing vision of urban co-creation, one that understands "produced environments [as] specific historical results of socio-environmental processes." Futurecraft is one of the engines by which social processes can become spatial products. It is the mechanism that Lefebvre polemically called for in 1972. In essence: "The democratization of [the right to the city], and the construction of a broad social movement to enforce its will, is imperative if the dispossessed are to take back the control which they have for so long been denied, and if they are to institute new modes of urbanization. Lefebvre was right to insist that the revolution has to be urban, in the broadest sense of that term, or nothing at all."[11]

With increasingly ubiquitous computing, urban life plays out at the convergence of physical and digital space, and a new citizenship may emerge. Forums for participation may allow the citizenry to engage in collective action and transform the space around them: the city of the future will emerge from collaborative futurecraft at the intersection of the digital and physical worlds.

ELEVEN

Using the lens of futurecraft, we have considered some of the key forces at play in the city today—from energy to building, from transportation to knowledge sharing. Each of these ultimately weaves into a tapestry of citizen empowerment, suggesting the possibility of human participation in operating (and even hacking) the city. Top-down frameworks, the kinds of systems in large multinational corporations, are not enough; bottom-up actions are needed to transform urban spaces. There can be no smart city without smart citizens.

As the city becomes a forum for action and reaction, the playing field of urban innovation will continue to reach new communities. By necessity, every technology begins at a particular place and time, and in many cases this is in research and development contexts in the industrially developed world. But from there it can expand in powerful and unpredictable ways. Twenty-five years ago, cell phones were exorbitantly expensive, a rarefied luxury for the global elite. Yet today, only a short time later, there are more active cell phones than there are human beings on the planet. In tandem with this diffusion are new opportunities for cell phones to serve as a tool of social empowerment.

Many of the technologies that are emerging in the developed world are likely to demonstrate their most potent impact in the entirely different contexts of emerging economies—where a myriad of challenges brought by urbanization will be decided in the decades to come. How might self-driving vehicles empower citizens and shape a new urban structure in countries that do not already have a robust public transportation network? How might digitally enhanced delivery networks help communities transform logistics in informal settlements? Across every sector, technological "leapfrogging" is becoming an increasingly important force in the world today.

Public engagement is a critical factor in discussions of our collective urban future. The vignettes and future scenarios throughout this book are not intended as predictions. They are ideas for debate, and we hope they will generate a critical conversation that could orient how we collectively think about transforming the present. How these technologies play out will depend on our own actions and reactions.

This book will be successful if it achieves the goal of sparking informed discussions. We do not intend it to be an agenda but a framework for actions that may have a transformative impact on cities worldwide. In the words of Buckminster Fuller: "We are called to be the architects of the future, not its victims."

NOTES

CHAPTER 1. FUTURECRAFT

Epigraph: L. Steven Sieden, *A Fuller View: Buckminster Fuller's Vision of Hope and Abundance for All* (Studio City, CA: Divine Arts, 2012), 101.

1. Thomas F. Anderson, "Boston at the End of the 20th Century," *Boston Globe*, December 24, 1900.

2. R. Buckminster Fuller and Kiyoshi Kuromiya, *Cosmography: A Posthumous Scenario for the Future of Humanity* (New York: Macmillan, 1992), 8.

3. Cellarius [Samuel Butler], "Darwin among the Machines," *The Press*, June 13, 1863.

4. George Basalla, *The Evolution of Technology* (Cambridge: Cambridge University Press, 1988).

5. Anthony Dunne and Fiona Raby, *Speculative Everything:*

Design, Fiction, and Social Dreaming (Cambridge, MA: MIT Press, 2013), 44.

6. Herbert A. Simon, *The Sciences of the Artificial* (Cambridge, MA: MIT Press, 1969), 114.

7. Cedric Price, *The Square Book* (London: Architectural Association Publications, 1984).

8. Lubomír Doležel, *Heterocosmica: Fiction and Possible Worlds* (Baltimore, MD: Johns Hopkins University Press, 1998), ix.

9. David Greelish, "An Interview with Computing Pioneer Alan Kay," *Time Magazine,* April 2, 2013.

CHAPTER 2. BITS AND ATOMS

Epigraph: Mark Weiser, "Ubiquitous Computing," *Ubicomp,* March 17, 1996, accessed June 30, 2015, http://www.ubiq.com/.

1. Marshall McLuhan and Gerald E. Stearn, eds., *McLuhan: Hot and Cool—A Primer for the Understanding of and a Critical Symposium with Responses by McLuhan* (New York: Dial, 1967), 279.

2. Manuel Castells, *The Rise of the Network Society,* vol. 1 of *The Information Age: Economy, Society and Culture* (Cambridge, MA: Blackwell, 1996), 412.

3. Nicholas Negroponte, *Being Digital* (New York: Alfred A. Knopf, 1995), 165.

4. Frances Cairncross, *The Death of Distance: How the Communications Revolution Will Change Our Lives* (Cambridge, MA: Harvard Business Press, 1997), 76.

5. Lord Richard Rogers, "Sustainable City, Lecture 1: The Culture of Cities," *Reith Lectures,* BBC Radio 4, February 12, 1995.

6. "The Great Sprawl of China," *The Economist,* January 24, 2015.

7. William J. Mitchell, *E-topia: "Urban Life, Jim—But Not as We Know It"* (Cambridge, MA: MIT Press, 1999), 76; F. Calabrese, Z. Smoreda, V. D. Blondel, and C. Ratti, "Interplay between Telecommunications and Face-to-Face Interactions: A Study Using Mobile Phone Data," *PLoS ONE* 6, no. 7 (2011).

8. Mitchell, *E-topia*, 3, 8.

9. Le Corbusier, *Vers une Architecture (Toward an Architecture)*, trans. John Goodman (Los Angeles: Getty Research Institute, 2007).

10. Carlo Ratti and Anthony Townsend, "The Best Way to Harness a City's Potential for Creativity and Innovation Is to Jack People into the Network and Get Out of the Way," *Scientific American*, September 2011, 42–48.

11. National Research Council, *Embedded, Everywhere: A Research Agenda for Networked Systems of Embedded Computers* (Washington, DC: National Academy Press, 2001), x.

12. Christopher Kelty, *Two Bits: The Cultural Significance of Free Software* (Durham, NC: Duke University Press Books, 2008), 2.

CHAPTER 3. WIKI CITY

Epigraph: Edward Glaeser, *Triumph of the City: How Our Greatest Invention Makes Us Richer, Smarter, Greener, Healthier, and Happier* (New York: Penguin Books, 2012).

1. Le Corbusier, *Ville Contemporaine de 3 Millions d'Habitants*, installation, Salon d'Automne, Paris, 1922.

2. Anthony Townsend, *Smart Cities: Big Data, Civic Hackers, and the Quest for a New Utopia* (New York: W. W. Norton, 2013), 28.

3. Pamela Licalzi O'Connell, "Korea's High-Tech Utopia, Where Everything Is Observed," *New York Times*, October 5, 2005.

4. Nikola Tesla, interview by John B. Kennedy, "When a Woman Is Boss," *Collier's Magazine,* January 30, 1926, 167.

5. Norbert Wiener, *The Human Use of Human Beings: Cybernetics and Society* (London: Houghton Mifflin, 1950), 17.

6. Mark Weiser, "The Computer for the 21st Century," *Communications, Computers, and Networks,* special issue of *Scientific American,* September 1991, 66–75.

7. Neil Gershenfeld, Raffi Krikorian, and Danny Cohen, "The Internet of Things," *Scientific American,* October 2004, 76–81.

8. Adam Greenfield, *Everyware: The Dawning Age of Ubiquitous Computing* (New York: New Riders, 2006).

9. Saskia Sassen, "The Global City: Introducing a Concept," *Brown Journal of World Affairs* 11.2 (2005): 27–43.

10. Rich Gold, "How Smart Does Your Bed Have to Be, Before You Are Afraid to Go to Sleep at Night?" *Cybernetics and Systems: An International Journal* 26.4 (1995): 379–386.

11. David Cameron, "Prime Minister's Statement on Disorder in England," Prime Minister's Office, 10 Downing Street, August 11, 2011.

12. Riot Cleanup, http://riotcleanup.co.uk (site suspended).

13. Saskia Sassen, "Big Data | Bad Data—an Open Forum," Engaging Data 2013, Senseable City Lab, MIT, Cambridge, MA, November 15, 2013.

14. Jane Jacobs, *The Death and Life of Great American Cities* (1961; New York: Vintage Books, 1992), 3.

15. Richard Sennett, "The Stupefying Smart City," LSE Cities, London School of Economics, December 2012; Carlo Ratti and Anthony Townsend, "Harnessing Residents' Electronic Devices Will Yield Truly Smart Cities," *Scientific American,* September 2011.

16. New Urban Mechanics, http://newurbanmechanics.org.

17. Catherine Tumber, "Unreal Cities?" *The Nation,* February 3, 2014, 35–37.

CHAPTER 4. BIG (URBAN) DATA

Epigraph: Italo Calvino, "World Memory," in *Numbers in the Dark,* trans. Tim Parks (New York: Vintage, 1995), 135.

1. Ibid.

2. The term "total recall" was coined by Gordon Bell: see Gordon Bell and Jim Gemmell, *Your Life, Uploaded: The Digital Way to Better Memory, Health, and Productivity* (New York: PLUME, 2009).

3. Bill Gates, foreword to *Your Life, Uploaded: The Digital Way to Better Memory, Health, and Productivity,* by Gordon Bell and Jim Gemmell (New York: PLUME, 2009), x–xi.

4. Eric Schmidt, "Google," Techonomy Conference, Lake Tahoe, CA, August 4, 2010, presentation at a panel discussion with Debby Hopkins, Kevin Kelly, and Lisa Randall, moderated by David Kirkpatrick.

5. S. Sobolevsky, I. Sitko, R. Tachet des Combes, B. Hawelka, J. M. Arias, and C. Ratti, "Money on the Move: Big Data of Bank Card Transactions as the New Proxy for Human Mobility Patterns and Regional Delineation—The Case of Residents and Foreign Visitors in Spain," *2014 IEEE International Congress on Big Data,* 136–143, http://www.ieee.org/conferences_events/.

6. F. Calabrese and C. Ratti, "Real Time Rome," *Networks and Communications Studies* 20 (2006): 247–258.

7. C. Kang, S. Sobolevsky, Y. Liu, and C. Ratti, "Exploring Human Movements in Singapore: A Comparative Analysis Based on Mo-

bile Phone and Taxicab Usages," *UrbComp '13: Proceedings of the 2nd ACM SIGKDD International Workshop on Urban Computing* (New York: ACM, 2013), article 1, http://dl.acm.org/citation.cfm?id=2505826; C. Ratti and K. Kloeckl, "Enacting the Real Time City, *Proceedings of Futur en Seine 2009*, Cap Digital, 2010, 72–84.

8. Carlo Ratti et al., "Investigation of the Waste-Removal Chain through Pervasive Computing," *IEEE Xplore: IBM Journal of Research and Development* 55.1.2 (2011): 1–11.

9. Kristofer Pister, "Emerging Challenges: Mobile Networking for 'Smart Dust.'" *Journal of Communications and Networks* 2.3 (2000): 188–196.

10. Bell and Gemmell, *Your Life, Uploaded.*

11. F. Girardin, F. Calabrese, F. Dal Fiore, C. Ratti, and J. Blat, "Digital Footprinting: Uncovering Tourists with User-Generated Content," *IEEE Pervasive Computing* 7.4 (2008): 36–43; F. Girardin, F. Dal Fiore, C. Ratti, and J. Blat, "Leveraging Explicitly Disclosed Location Information to Understand Tourist Dynamics: A Case Study," *Journal of Location-Based Services* 2.1 (2008): 41–54.

12. Adam D. I. Kramer, Jamie E. Guillory, and Jeffrey T. Hancock, "Experimental Evidence of Massive-Scale Emotional Contagion through Social Networks," *Proceedings of the National Academy of Sciences* 111.24 (2014): 8788–8790.

13. Carlo Ratti and Otto Ng, "One Country, Two Lungs," MIT Senseable City Lab with LAAB, exhibition, presented at the Hong Kong and Shenzhen Bi-City Biennale of Urbanism and Architecture, 2013.

14. Hamed Haddadi, Heidi Howard, Amir Chaudhry, Jon Crowcroft, Anil Madhavapeddy, and Richard Mortier, "Personal Data:

Thinking Inside the Box," arXiv.org [cs.CY], 2015, http://arxiv
.org/abs/1501.04737.

CHAPTER 5. CYBORG SOCIETY

Epigraph: Amber Case, "We Are All Cyborgs Now," *TEDWomen*,
International Trade Center, Washington, DC, December 8, 2010.

1. Manfred Clynes and Nathan Kline, "Cyborgs and Space," *Astro-nautics*, September 1960, 26–27, 74–76.

2. Ariane Lourie Harrison, introduction to *Architectural Theories of the Environment: Posthuman Territory*, ed. Harrison (New York: Routledge, 2013), 3–35.

3. André Leroi-Gourhan, *Le Geste et la Parole*, 2 vols. (Paris: Albin Michel, 1964–1965), published in English as *Gesture and Speech*, trans. Anna Bostock Berger (Cambridge, MA: MIT Press, 1993).

4. Antoine Picon, "Architecture and the Virtual: Towards a New Materiality," *Praxis: New Technologies New Architectures* 6 (2004): 114–121.

5. Donna Haraway, "A Cyborg Manifesto: Science, Technology, and Socialist-Feminism in the Late Twentieth Century," in *Simians, Cyborgs and Women: The Reinvention of Nature*, ed. Haraway (New York: Routledge, 1991), 149–181; Paul Virilio, "The Law of Proximity," in *Book for the Unstable Media*, ed. Alex Adriaansens, Joke Brouwer, Rik Delhaas, and Eugenie den Uyl (Rotterdam: V2_, 1992).

6. Virilio, "Law of Proximity."

7. Francis Fukuyama, *Our Posthuman Future: Consequences of the Biotechnology Revolution* (New York: Picador, 2002); Ray Kurzweil, *The Singularity Is Near: When Humans Transcend*

Biology (New York: Viking, 2005); William Mitchell, *Me++: The Cyborg Self and the Networked City* (Cambridge, MA: MIT Press, 2004).

8. Toyo Ito, "Mediatheque of Sendai," in *Toyo Ito,* ed. Ron Witte and Hiroto Kobayashi (Munich: Prestel; Cambridge: Harvard University Graduate School of Design, 2002).

9. Bruno Latour, Valérie November, and Eduardo Camacho-Hübner, "Entering Risky Territory: Space in the Age of Digital Navigation," *Environment and Planning D: Society and Space* 28 (2010): 581–599.

10. Case, "We Are All Cyborgs Now."

11. Antoine Picon, "La Ville Territoire des Cyborgs," *Flux* 15.36–37 (1999): 76–79.

12. Matthew Gandy, "Cyborg Urbanization: Complexity and Monstrosity in the Contemporary City," *International Journal of Urban and Regional Research* 29.1 (2005): 28.

13. Picon, "Architecture and the Virtual."

14. Martijn de Waal, "The Ideas and Ideals in Urban Media Theory," in *From Social Butterfly to Engaged Citizen: Urban Informatics, Social Media, Ubiquitous Computing, and Mobile Technology to Support Citizen Engagement,* ed. Marcus Foth, Laura Forlano, Christine Satchell, and Martin Gibbs (Cambridge, MA: MIT Press, 2011), 5–20.

15. Caroline McCarthy, "Dodgeball: A Eulogy," *CNet,* January 15, 2009, http://www.cnet.com.

16. Picon, "Architecture and the Virtual."

17. Howard Rheingold, *Smart Mobs: The Next Social Revolution* (New York: Basic Books, 2002).

18. Michael Chorost, *Rebuilt: How Becoming Part Computer Made Me More Human* (New York: Houghton Mifflin, 2005).

19. Vernor Vinge, "First Word," *Omni Magazine,* January 1983.

CHAPTER 6. LIVING ARCHITECTURE

Epigraph: *Frank Lloyd Wright: An Autobiography* (Petaluma, CA: Pomegranate Communications, 2005), 149.

1. Le Corbusier and María Cecilia O'Byrne Orozco, "Elaboration du Plan Régulateur de Bogotá," in *Le Corbusier en Bogotá,* ed. María Cecilia O'Byrne Orozco (Bogotá: Universidad de Los Andes, Facultad de Arquitectura y Diseño, 2010).

2. Patrick Schumacher, "Parametricism as Style—Parametricist Manifesto," 11th Architecture Biennale, Dark Side Club, Venice, Italy, November 2008.

3. Kurt Forster, *Metamorph: Catalogue for the Venice Biennale of Architecture* (New York: Rizzoli, 2004); "9th International Architecture Exhibition in Venice (Part 1)," *designboom,* 2004, acesssed June 22, 2015, http://www.designboom.com.

4. Alejandro Zaera-Polo, "The Hokusai Wave," *Volume 3,* 2005.

5. Stan Allen, "From the Biological to the Geological," in *Landform Building: Architecture's New Terrain* (Zurich: Lars Muller Publishers, 2011).

6. Antoine Picon, "Digital/Minimal?" *Architettura,* February 25, 2006.

7. Ibid.

8. Gordon Pask, "The Architectural Relevance of Cybernetics," *Architectural Design,* September 1969, 494–496.

9. Ibid.

10. Paola Antonelli, *Design and the Elastic Mind,* exhibition catalogue (New York: Museum of Modern Art, 2008).

11. Ibid.

12. Anthony Dunne and Fiona Raby, *Speculative Everything: Design, Fiction, and Social Dreaming* (Cambridge, MA: MIT Press, 2013).

13. William J. Mitchell, *City of Bits: Space, Place, and the Infobahn* (Cambridge, MA: MIT Press, 1996), 43, https://mitpress.mit.edu/books/city-bits.

14. Picon, "Digital/Minimal?"

15. Allen, "From the Biological to the Geological"; Matthew Gandy, "Cyborg Urbanization: Complexity and Monstrosity in the Contemporary City," *International Journal of Urban and Regional Research* 29.1 (2005): 29.

16. Erik Swyngedouw, "Circulations and Metabolisms: (Hybrid) Natures and (Cyborg) Cities," *Science as Culture, Special Issue: Technonatural Time-Spaces* 15.2 (2006): 105–121.

CHAPTER 7. MOBILITY

Epigraph: Lewis Mumford, *My Works and Days: A Personal Chronicle* (New York: Houghton Mifflin Harcourt Press, 1979).

1. Ibid.

2. Thomas Sugrue, "From Motor City to Motor Metropolis: How the Automobile Industry Reshaped Urban America," *Automobile in American Life and Society,* University of Michigan-Dearborn and Benson Ford Research Center, 2004, accessed June 22, 2015, http://www.autolife.umd.umich.edu/.

3. Jane Jacobs, *Dark Age Ahead* (2003; reprint, Knopf Doubleday, 2007), 187–188.

4. Reyner Banham, *Los Angeles: The Architecture of Four Ecologies* (New York: Harper and Row, 1971). The documentary is "Reyner Banham Loves Los Angeles," *One Pair of Eyes,* BBC Films, 1972.

5. Anthony Downs, "The Law of Peak-Hour Expressway Congestion," *Traffic Quarterly* 16.3 (1962): 393–409.

6. Lewis Mumford, *The Highway and the City* (New York: Harcourt, Brace and World, 1957), 238.

7. "7 Million Premature Deaths Annually Linked to Air Pollution," World Health Organization, March 25, 2014, accessed June 22, 2015, http://www.who.int.

8. Donald Shoup, "The High Cost of Free Parking," *Journal of Planning Education and Research* 17 (1997): 3–20.

9. Ibid.

10. J. Firnkorn and M. Müller, "Selling Mobility Instead of Cars: New Business Strategies of Automakers and the Impact on Private Vehicle Holding," *Business Strategy and the Environment* 21 (2012): 264–280; Elliot Martin, Susan Shaheen, and Jeffrey Lidicker, "Impact of Carsharing on Household Vehicle Holdings: Results from a North American Shared-Use Vehicle Survey," *Transportation Research Record: Journal of the Transportation Research Board* 2143 (2010): 150–158.

11. Paolo Santi and Carlo Ratti, "Quantifying the Benefits of Vehicle Pooling with Shareability Networks," *PNAS* 111.37 (2014): 13290–13294.

12. Ibid.

13. Brandon Schoettle and Michael Sivak, "The Reasons for the Recent Decline in Young Driver Licensing in the U.S.," 2013,

University of Michigan Transportation Research Institute, http://hdl.handle.net/2027.42/99124.

CHAPTER 8. ENERGY

Epigraph: William J. Mitchell, *E-topia: "Urban Life, Jim—But Not as We Know It"* (Cambridge, MA: MIT Press, 1999), 5.

1. Ibid.
2. Reyner Banham and François Dallegret, "A Home Is Not a House," *Art in America* 2 (April 1965): 70–79.
3. C. Martani, D. Lee, P. Robinson, R. E. Britter, and Carlo Ratti, "ENERNET: Studying the Dynamic Relationship between Building Occupancy and Energy Consumption," *Energy and Buildings* 47 (2012): 584–591.
4. Ibid.
5. Buckminster Fuller, *Critical Path* (New York: St. Martin's, 1981), 206.
6. Ibid.
7. Karl Johan Åström and Richard M. Murray, *Feedback Systems* (Princeton, NJ: Princeton University Press, 2008), 1.
8. U.S. Department of Energy, Office of Electricity Delivery and Energy Reliability, "What Is the Smart Grid?" SmartGrid.gov, accessed June 22, 2015, http://www.smartgrid.gov/.

CHAPTER 9. KNOWLEDGE

Epigraph: Paul Markillie, "A Third Industrial Revolution," *The Economist*, April 21, 2012.

1. Thorstein Veblen, *The Theory of the Leisure Class: An Economic Study of Institutions* (New York: B. W. Heubsch, 1899).

2. Lewis Mumford, "The Fulfillment of Man," in *The Conduct of Life*, ed. Mumford (New York: Harcourt, Brace, 1951).

3. "The Automobile in Le Corbusier's Ideal Cities" (PDF file, MIT Press, Cambridge, MA).

4. Sigfried Giedion, *Mechanization Takes Command: A Contribution to Anonymous History* (Minneapolis: University of Minnesota Press, 2014), 77.

5. Johan Huizinga, *Homo Ludens: A Study of the Play-Element in Culture*, trans. Johan Huizinga and R. F. C Hull (London: Routledge and Kegan Paul, 1944).

6. Constant Nieuwenhuys, *New Babylon*, exhibition catalogue (The Hague: Municipal Museum, 1974).

7. Markillie, "Third Industrial Revolution."

8. David Benjamin, interview by Jessica Liss, "The Living," *Emerging Voices 2014*, Architectural League of New York, 2014, accessed June 22, 2015, http://archleague.org/2014/07/the-living/.

9. Neil Gershenfeld. "Unleash Your Creativity in a Fab Lab," TED Conference, Portola Plaza Hotel, Monterey, CA, February 23, 2006.

10. Ibid.

11. William J. Mitchell, *E-topia: "Urban life, Jim—But Not as We Know It"* (Cambridge: MIT Press, 1999), 3.

12. Ibid.

CHAPTER 10. HACK THE CITY

Epigraph: Ben Lillie, "Can a City Be Too Technological? Saskia Sassen at TED2013," *TEDBlog*, February 27, 2013, accessed June 30, 2015, http//blog.ted.com/.

1. *Merriam-Webster Dictionary* (2015), http://www.merriam-webster .com/, s.v. "hacker."

2. Mark Weiser, "Ubiquitous Computing," *Ubicomp,* March 17, 1996, accessed June 30, 2015, http://www.ubiq.com/.

3. Richard Stallman, "On Hacking," *Richard Stallman's Personal Site,* 2014, accessed June 30, 2015, https://stallman.org.

4. Robert Moore, *Cybercrime: Investigating High-Technology Computer Crime* (New York: Routledge; Newark, NJ: LexisNexis/ Matthew Bender, 2005).

5. Kevin Poulsen, "Five Years behind Bars for Hacking Wasn't Punishment Enough: Meet the Amazing Modemless Man," *Wired Magazine,* 2003, accessed June 30, 2015, http://www.wired.com.

6. Tech Model Railroad Club, "A Brief History of the Tech Model Railroad Club," *Tech Model Railroad Club of MIT,* Massachusetts Institute of Technology, 2015, accessed June 30, 2015, http:// tmrc.mit.edu.

7. David Kushner, "The Real Story of Stuxnet," *IEEE Spectrum,* 2015, accessed June 30, 2015, http//spectrum.ieee.org.

8. Ibid.

9. Saskia Sassen, interview by Open Source Urbanism, "Saskia Sassen," *OSU// The Interviews,* Open Source Urbanism, November 2013, accessed June 30, 2015, https://opensourceurbanism. wordpress.com; Henri Lefebvre, *Writings on Cities,* trans. and ed. Eleonore Kofman and Elizabeth Lebas (Oxford, UK: Wiley-Blackwell, 1996), 158.

10. Lefebvre, *Writings on Cities,* 158; Erik Swyngedouw, "Circulations and Metabolisms: (Hybrid) Natures and (Cyborg) Cities," *Science*

as Culture, Special Issue: Technonatural Time-Spaces 15.2 (2006): 105–121.

11. Swyngedouw, "Circulations and Metabolisms"; David Harvey, "The Right to the City," *New Left Review* 53 (September–October 2008): 23–40.

ACKNOWLEDGMENTS

The kernel of this book developed over several years of research through the Senseable City Laboratory, at the Massachusetts Institute of Technology. Constructing a conceptual framework for this work led us to the idea that we call futurecraft. It serves as a tool to reflect on our past projects and as a guide for future efforts in urban practice. Futurecraft has now become a familiar concept to the researchers, students, and staff who spend their time in our crowded corner of MIT's Building 9: it's just how we think about research and design in urban space. Without this talented group of people—and in particular Associate Director Assaf Biderman—this book would have not been possible.

As we started to work on this book, and to critically reflect on our past projects and the methodology that informed them, it became clear that opening the process to outside discussion would enrich our critical endeavor. In the true spirit of future-craft, we began this investigation with a seminar at MIT, titled "The Future of the City" (MIT 11.318), during the spring semester of 2014. The students' analytical writing, discussion, and debate were a crucial part of developing this work. We consider each member of the class—Alicia Rouault, Benjamin Scheerbarth, Dima Rachid, Jacob Koch, and Sandra Poizat—an integral contributor.

Beyond the class, we are indebted to the fertile environment of MIT. We would like to thank our colleagues for their great work, intellectual engagement, and visionary attitude—in particular, Dennis Frenchman, Eran Ben-Joseph, Frank Levy, Hiroshi Ishii, Larry Vale, and the late William Mitchell. More broadly, a number of scholars and practitioners outside MIT have inspired this book, directly or indirectly—above all, Michael Batty, Saskia Sassen, Geoffrey West, Antoine Picon, Anthony Townsend, and Andrea Bosco, the editor of our prior writing in Italian. We are also particularly indebted to Joe Calamia, our editor at Yale University Press, for his continuous patience, guidance, and encouragement.

Finally, we would like to thank you, the engaged reader and active citizen. We leave you with a challenge: reinvent your city!

CREDITS

The Evolution of Culture by Augustus Henry Lane-Fox
   Pitt-Rivers: Augustus Henry Lane-Fox Pitt-Rivers, *The Evo-*
   lution of Culture (Oxford, 1906; reprinted, New York: AMS
   Press, 1979), pl. III.

New York Talk Exchange by the MIT Senseable City Laboratory
Boston 311 Map by the MIT Senseable City Laboratory
Real Time Rome by the MIT Senseable City Laboratory
Trash Track by the MIT Senseable City Laboratory
Ojos del Mundo by the MIT Senseable City Laboratory
HubCab by the MIT Senseable City Laboratory
Local Warming by the MIT Senseable City Laboratory
   Full project credits are at http://senseable.mit.edu.

Formula One Telemetry of Adrian Sutil: Creative Commons,
   https://www.flickr.com/photos/sgmendez/3524265009/.
   Photograph by Salva Mendez.

Songdo Central Park: Creative Commons, https://www.flickr.com/photos/traveloriented/13308944185/. Photograph by Travel Oriented.

Spider Mite on Mirror Assembly by Sandia National Laboratories: Courtesy Sandia National Laboratories, SUMMiT™ Technologies, www.sandia.gov/mstc.

Guggenheim Museum Bilbao by Frank Gehry: Creative Commons, https://commons.wikimedia.org/wiki/File:Guggenheim_Bilbao_may-2006.jpg. Photograph courtesy of Samuel Negredo.

Digital Water Pavilion by Carlo Ratti Associati
Vertical Plotter by Carlo Ratti Associati
Photographs from Carlo Ratti Associati.

Plan of Brasília by Oscar Niemeyer and Lúcio Costa: Digital photo of the original plan, on display at the O Espaço Lúcio Costa in Brasília. Creative Commons, https://commons.wikimedia.org/wiki/File:Brasilia_-_Plan.jpg. Photograph by Uri Rosenheck, desaturated by Tetraktys.

Copenhagen Wheel by the MIT Senseable City Laboratory and Superpedestrian: The project originated at the Senseable City Lab and has been developed and commercialized at Superpedestrian.

Past and Future of the Thermostat: Examples from Nest and Honeywell Labs. Photographs courtesy of Nest and Honeywell.

Royal Saltworks at Arc-et-Senans by Claude Nicolas Ledoux, 1775–79: Creative Commons, https://commons.wikimedia.org/wiki/

File:Carte_g%C3%A9n%C3%A9rale_des_environs_de_la_
Saline_de_Chaux.jpg. Photograph by Jean-Christophe Benoist.

Fab Lab by Applied Nomadology: Creative Commons, https://www
.flickr.com/browser/upgrade/?continue=/photos/centralasian/
5044869128/.

People Congregated in Tahrir Square on February 9, 2011, by Jonathan Rashad: Wikimedia Commons, https://en.wikipedia.org/
wiki/Tahrir_Square#/media/File:Tahrir_Square_-_February_9,
_2011.png. Photograph courtesy of the photographer.

Pink Condom by Oliviero Toscani for United Colors of Benetton,
December 1, 1993: Photograph courtesy of the artist.